Bridge Analysis
by Microcomputer

The McGraw-Hill Infrastructure Series

PLANNING

DICKEY   *Metropolitan Transportation Planning, 2/e (1983)*
HORONJEFF & McKELVEY   *Planning and Design of Airports, 3/e (1983)*
HUTCHINSON   *Principles of Urban Transport Planning (1974)*
KANAFANI   *Transportation Demand Analysis (1983)*
MORLOCK   *Introduction to Transportation Engineering and Planning (1978)*

ANALYSIS & INSPECTION

BAKHT & JAEGER   *Bridge Analysis Simplified (1985)*
BOWLES   *Foundation Analysis and Design (1982)*
LAURSEN   *Structural Analysis (1978)*
LEET   *Reinforced Concrete Design (1984)*
NAAMAN   *Prestressed Concrete Analysis and Design (1982)*
RAU & WOOTEN   *Environmental Impact Analysis Handbook (1979)*
SACK   *Structural Analysis (1984)*
WILLIAMS & LUCAS   *Structural Analysis for Engineers (1967)*
WINTER & NILSON   *Design of Concrete Structures, 9/e (1979)*

ENGINEERING & DESIGN

BILLINGTON   *Thin-Shell Concrete Structures, 2/e (1982)*
BRUSH & ALMROTH   *Buckling of Bars, Plates, and Shells (1975)*
FULLER   *Engineering of Pile Installations (1983)*
GAYLORD & GAYLORD   *Structural Engineering Handbook, 2/e (1979)*
HALL & DRACUP   *Water Resource Systems Engineering (1970)*
LePATNER & JOHNSON   *Structural and Foundation Failures: A Casebook for Architects, Engineers, and Lawyers (1982)*
KREBS & WALKER   *Highway Materials (1971)*
QUINN   *Design and Construction of Ports and Marine Structures (1972)*
RINALDI   *Modeling and Control of River Quality (1978)*
ROBINSON   *Highways and Our Environment (1971)*
WALLACE & MARTIN   *Asphalt Pavement Engineering (1967)*
YANG   *Design of Functional Pavements (1973)*

For more information about other McGraw-Hill materials,
call 1-800-2-MCGRAW in the United States. In other
countries, call your nearest McGraw-Hill office.

Bridge Analysis by Microcomputer

Leslie G. Jaeger
*Research Professor of Civil Engineering
and Applied Mathematics,
The Technical University of Nova Scotia*

Baidar Bakht
*Principal Research Engineer,
Ministry of Transportation of Ontario*

McGraw-Hill Book Company

New York   St. Louis   San Francisco   Auckland   Bogotá
Caracas   Colorado Springs   Hamburg   Lisbon
London   Madrid   Mexico   Milan   Montreal
New Delhi   Oklahoma City   Panama   Paris
San Juan   São Paulo   Singapore
Sydney   Tokyo   Toronto

Library of Congress Cataloging-in-Publication Data

Jaeger, Leslie G.
   Bridge analysis by microcomputer / Leslie G. Jaeger, Baidar Bakht.
      p.   cm.
   Includes index.
   ISBN 0-07-032333-X
   1. Bridges—Design—Data processing.   2. Microcomputers.
I. Bakht, Baidar   II. Title.
 TG300.B33 1989
 624′.25—dc19

1234567890   DOC/DOC   8954321098

ISBN 0-07-032333-X

*The editors for this book were Nadine Post and Beatrice E. Eckes, the
designer was Naomi Auerbach, and the production supervisor was
Suzanne Babeuf. It was set in Century Schoolbook by Techna Type.*

Printed and bound by R. R. Donnelley & Sons Company.

*For more information about other McGraw-Hill materials,
call 1-800-2-MCGRAW in the United States. In other
countries, call your nearest McGraw-Hill office.*

For our children,
Valerie and Hilary Jaeger
and Natasha and Sacha Bakht.

Contents

Preface

Methods of transverse load distribution analysis of highway bridges range in sophistication from the overly simplified American Association of State Highway and Transportation Officials (AASHTO) methods to highly complex finite element methods. The former, because of being too simple, are excessively conservative, and the latter, which require fairly complex standard programs, are prone to common errors of idealization and interpretation of results as well as being relatively costly.

This book is a companion volume to *Bridge Analysis Simplified* (McGraw-Hill, 1985). The earlier book expands upon the simple AASHTO approach to provide a number of more accurate manual methods of analysis for application in conjunction with North American and similar codes. Our objective in writing this second work is to provide to bridge designers alternative methods of bridge analysis which are refined, which can be implemented through even the most basic microcomputer, and which do not have the limitations of the simplified methods.

In the present work, a refined method known as the *semicontinuum method* is developed which requires only a small microcomputer while retaining the accuracy and versatility of other refined methods that usually require large mainframe computers. The semicontinuum method not only provides a considerable saving in the time required to perform the analysis but also does not remove the designer from the physical "feel" of the structure.

The book functions at three levels:

1. For those who have a liking for the mathematical kind of structural analysis, it provides the historical background and step-by-step derivations for a thorough understanding. This treatment is valuable, for example, for graduate students.

2. Instructions are provided for writing computer programs for those who want to write their own programs themselves but are not very much concerned with the mathematical derivations on which the pro-

grams are based. In some circumstances it may be advantageous to the engineer to write his or her own programs in preference to using those given in the text. As an example, building into the program details of a particular design loading, such as AASHTO HS 20, rather than inputting the individual loads and load positions as data, may be very valuable. The semicontinuum method presents a simple-to-use and easy-to-idealize alternative; programs utilizing this method can be written by any engineer in a fraction of the time needed to develop, for example, a grillage analogy program.

3. Ready-to-use computer programs and also two manual methods are given for those who, having satisfied themselves of the accuracy of the methods, wish to implement them directly in the design office. The book ensures that even at this level the user of the program does not lose touch with the physical significance of the bridge behavior.

Most bridge design codes, including AASHTO, approve *rational* methods of analysis. The semicontinuum method, being a rational method whose accuracy is similar to that of the already well-tested and approved grillage analogy method, clearly falls into this category. Since local jurisdictions within a country have their own interpretations of the national codes of practice, an engineer using this book to design a bridge for a particular locale should be aware of that locale's interpretations of the relevant national code.

It is noted that, unlike *Bridge Analysis Simplified,* this book is not limited to AASHTO, Ontario, and similar highway bridge codes. The methods presented are general enough to deal with any loading and with any structural complications, such as unequal spacing of longitudinal girders, if present. The methods can also account for any degree of torsional rigidities in both the longitudinal and the transverse directions.

Leslie G. Jaeger
Baidar Bakht

Acknowledgments

The work that led to this book was partly carried out under research funding provided by the Natural Sciences and Engineering Research Council of Canada and partly at the Ministry of Transportation of Ontario. The support of these two organizations is gratefully acknowledged.

We are also grateful to a number of individuals including our research colleagues Dr. Roger Dorton, Dr. Aftab Mufti, Mr. Peter Smith, and Dr. Chandra Surana for supporting our endeavors in a number of ways; to Mrs. Maria Feehan for typing the entire manuscript of the book; to Mrs. Roberta Sturge for typing many of the papers that were the forerunners of it; to Messrs. Rob Lockhart, John Nyers, and Periya Naidu for their help in creating the artwork for the original papers; and to all those students and technicians who helped us in developing the various computer programs.

A large portion of the book is based upon papers written by us and published in the *Journal of the Structural Division of the American Society of Civil Engineers*, the *Canadian Journal of Civil Engineering*, the *Journal of the Institution of Engineers (India)*, and *The Bridge and Structural Engineer, India*. The permission of the editors of these journals to draw from our papers is gratefully acknowledged.

Leslie G. Jaeger
Baidar Bakht

Bridge Analysis
by Microcomputer

Introductory Concepts

1.1 Analysis

The word *analysis* implies the conceptual breaking up of a whole into parts so that one can have an insight into the complete entity. In the context of structural engineering, analysis usually refers to *force analysis,* a process in which one determines the distribution of force effects or responses, such as deflections and bending moments, in the various components of the structure. Another less commonly used term in structural engineering is *strength analysis,* which refers to the process of determining the strength of the whole structure or its components. The term *analysis* is used in this book only in the meaning of force analysis. In particular, it refers to the determination of bending moments, twisting moments, shear forces, and deflections in the main components of the superstructures of common bridge types.

1.1.1 Load distribution

In the context of bridges, the term *load distribution* is often used instead of analysis. It may be instructive to establish the meaning of this term.

As a vehicle for explanation, one considers three simply supported beams of span L placed side by side. As shown in Fig. 1.1, the middle beam carries a central-point load P. In the absence of any connection between these beams, the entire load will, of course, be sustained by the externally loaded beam and the bending-moment diagram will be the familiar triangular one, as shown in Fig. 1.1, with the maximum value of $PL/4$.

If the three longitudinal beams are now connected by a transverse beam at midspan, then, as shown in Fig. 1.2, a portion of the load P will be taken by the outer beams, which do not themselves carry the external load. How much of the externally applied load is distributed to the outer beams will depend upon the beam rigidities and their spans and spacings. If the flexural rigidity of the transverse beam is a very small fraction of the rigidities of longitudinal beams and the spacings of longitudinal beams are relatively large, then hardly any load will be transferred to the beams not carrying the direct load. On the other hand, an infinitely stiff transverse beam will compel the three beams to deflect equally, and in that case each beam will accept a load of $P/3$.

If the load accepted by each of the symmetrical outer beams is P_2, then the bending-moment diagrams for the three beams will be as shown in Fig. 1.2.

Clearly, the transference of load from the externally loaded beam to the beams not carrying the direct load takes place because the loaded beam has a tendency to deflect more than its neighbors and the tendency of the transverse beam is to reduce the differential deflection between adjacent beams. As is discussed in Sec. 1.2, the assembly of longitudinal and transverse beams can be mathematical models of a slab-on-girder bridge with the longitudinal beams representing the girders and one or more transverse beams representing the deck slab. The usual notation for the two directions is illustrated in Fig. 1.3. The process of load transference from the loaded girders to girders which are not directly loaded is often referred to as *load distribution*. If an externally loaded girder in a bridge retains most of the applied loads, with only a small fraction of load being transferred to other girders, the bridge is said to have poor distribution characteristics. Bridges

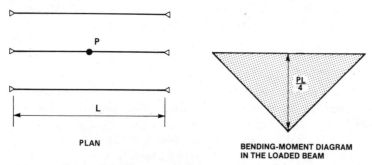

PLAN

BENDING-MOMENT DIAGRAM
IN THE LOADED BEAM

Figure 1.1   The case of unconnected beams.

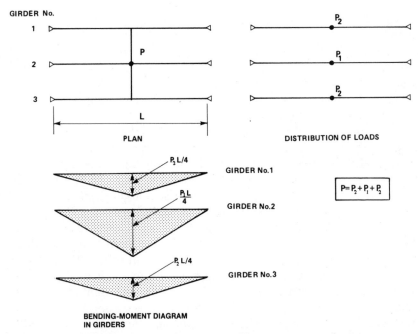

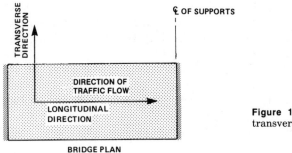

Figure 1.2 Load distribution in girders connected by one transverse beam.

with good distribution characteristics are clearly those in which the applied loading is well distributed between the various girders.

1.1.2 Distribution factors and coefficients

Distribution factors (DFs) and distribution coefficients are nondimensional measures of load distribution in a right bridge. At a given point of a transverse section the DF for a certain response is equal to the

Figure 1.3 Longitudinal and transverse directions.

ratio of the response at that point to the average response at the cross section containing the point. For example, in the case shown in Fig. 1.2, the average bending moment in the three girders is $PL/12$, so that the DF for longitudinal moment in the middle beam at midspan is equal to $(P_1L/4)/(PL/12)$, or $3P_1/P$. It is noted that the terms *distribution factor* and *distribution coefficient* are interchangeable. The latter term is mostly used in the book.

In the very simple case shown in Fig. 1.2 the bending-moment diagrams of the three girders are all of the same shape, being the familiar triangle. The bending-moment diagram for any one girder can then be regarded as the *free-bending-moment diagram,* shown in Fig. 1.1, multiplied by a number which is called the *distribution coefficient.* Referring again to Fig. 1.2, the distribution coefficients are P_2/P, P_1/P, P_2/P. Simplified methods of bridge analysis (see below) are based upon the assumption that all girders of the bridge receive net loadings which are similar to one another and are obtained by multiplying the external loading on the bridge by the relevant distribution coefficients. In these simplified methods, therefore, the distribution coefficients for one response, such as bending moment, are the same as for another response, such as shear force; furthermore, these distribution coefficients are the same at all longitudinal positions of the bridge, so that if one particular girder takes, say, one-fifth of the free bending moment at any one transverse cross section of the bridge, it is deemed to take that same fraction at all other transverse cross sections.

1.1.3 Simplified versus rigorous methods

The various methods of bridge analysis are frequently referred to in the literature as either *simplified* or *rigorous.* One may say intuitively that a simplified method requires fewer computations than a rigorous one; however, a clear distinction between the two is not self-evident. In what follows the distinction between simplified and rigorous methods of bridge analysis is given a precise meaning. As a vehicle for explanation, the example of Fig. 1.2 is used, but with the addition of transverse beams at the quarter-span and three-quarter-span points. This new assembly of beams is shown in Fig. 1.4; for simplicity, torsionless behavior is assumed.

Before the introduction of the additional transverse beams, the deflections of the middle beam at the quarter-span and three-quarter-span points are larger than the corresponding deflections in the two outer beams. The introduction of transverse beams at these points has the effect of reducing the differential deflection between the middle and outer beams, and in so doing an upward force ΔP is introduced at each of the two points in the middle beam. The outer beams each receive a downward force of $\Delta P/2$ at each of the points, as shown in Fig. 1.4. It is readily verified, as will be demonstrated later, that the introduction of the additional transverse beams also has the effect of slightly changing the previous values of the interactive forces P_1 and P_2.

As can be seen in Fig. 1.4, the introduction of the beams and the consequent appearance of reactive forces at the quarter-span and three-quarter-span points changes the shape of the bending-moment diagrams. Instead of being triangular as they were for the case shown in Fig. 1.2, the bending-moment diagrams now become polygonal; for the outer girders the polygon is convex downward, while for the center girder the polygon is concave downward in each half of the span. Thus, even in the case of only three longitudinal girders and only three transverse beams it is apparent that the fraction of free bending moment which is accepted by any one girder varies appreciably along the span.

In a real-life bridge the action of the deck slab can be represented

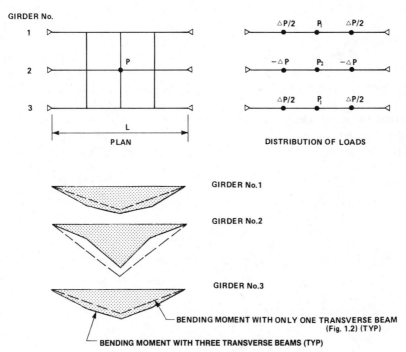

Figure 1.4   Load distribution in beams connected by three transverse beams.

Figure 1.5 Shape of bending-moment diagrams. (*a*) Bending-moment diagram for a directly loaded girder. (*b*) Bending-moment diagram for a girder not carrying a direct load.

by an infinite number of conceptual transverse beams. For such a case it can be readily shown that, for a single-point load at midspan, the bending-moment diagram for the loaded girder has the shape shown in Fig. 1.5*a* and for a girder not carrying a direct load the shape shown in Fig. 1.5*b*.

If the bending-moment diagrams for all girders have *exactly* the same shapes, then for any girder the ratio of the actual moment and the average moment remains constant along the span. Moreover, this ratio is the same as the ratio of actual and average shear forces. In other words, when the bending-moment diagrams for the various girders are of *exactly* the same shape, the DFs for girder moments are the same as those for girder shears and their values do not change along the span. It has just been shown that when two or more transverse members interconnect the longitudinal girders, then for most patterns of external load the bending-moment diagrams for the various girders do not have exactly similar shapes. In this case, the DF for moments is no longer equal to that for shears, and further its values do not remain constant along the span. To supplement the qualitative discussion given above, it is instructive to study this aspect of load distribution quantitatively with the help of a specific example. The example, which now follows, is solved with the help of the familiar beam deflection theory, ignoring torsional effects. The example is that of three equally spaced girders connected with three symmetrically placed transverse beams at the one-sixth-span and midspan positions. As shown in Fig.

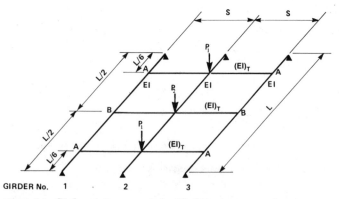

Figure 1.6 Girders interconnected with three transverse beams.

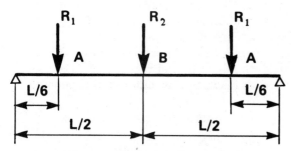

Figure 1.7 A beam under three loads.

1.6, the middle girder is subjected to three symmetrical externally applied loads P_1, P_2, and P_1 at the intersections of the girder and transverse beams.

To analyze this assembly of beams, the case of a simple girder subjected to three loads R_1, R_2, and R_1 is first studied. This case is shown in Fig. 1.7. It can be shown from simple beam theory that the deflections at points A and B, identified in Fig. 1.7 and denoted as w_A and w_B respectively, are given by the following expressions.

$$w_A = \frac{14R_1 + 13R_2}{1296EI} L^3 \tag{1.1a}$$

$$w_B = \frac{26R_1 + 27R_2}{1296EI} L^3 \tag{1.1b}$$

where the notation is as shown in Fig. 1.6.

Owing to load distribution, the outer girders are subjected to downward forces R_1, R_2, and R_1 at their intersections with the transverse beams and the middle girder is subjected to net point loads of $(P_1 - 2R_1)$, $(P_2 - 2R_2)$, and $(P_1 - 2R_1)$. These loads on individual girders are shown in Fig. 1.8. Deflections of the center girder at the points of

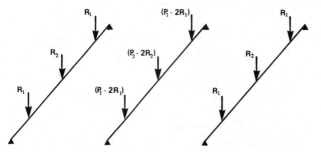

Figure 1.8 Net forces on individual girders.

load application can be readily obtained by using Eq. (1.1) by putting $(P_1 - 2R_1)$ in place of R_1 and $(P_2 - 2R_2)$ in place of R_2.

The transverse beams are subjected to the reactive forces shown in Fig. 1.9. For these loadings, the deflection of transverse beam AA at the middle girder position with respect to the outer ones is denoted by w_A and is given by

$$w_A = \frac{R_1 S^3}{3(EI)_T} \tag{1.2a}$$

Similarly, the relative midspan deflection w_B of transverse beam BB is given by

$$w_B = \frac{R_2 S^3}{3(EI)_T} \tag{1.2b}$$

From compatibility of deflections, it is obvious that w_A and w_B should be equal to the differences in central-girder and outer-girder deflections at sections AA and BB respectively. These equalities lead to two simultaneous equations by solving which the unknowns R_1 and R_2 can be readily found. For the particular case of $P_1 = 0$ and $P_2 = P$, the two reactive forces are found to have the following expressions:

$$R_1 = P \frac{26\eta'}{360\eta'^2} + 246\eta' + 4 \tag{1.3a}$$

$$R_2 = P \frac{\eta'(54 + 120\eta')}{360\eta'^2} + 246\eta' + 4 \tag{1.3b}$$

where η' is given by

$$\eta' = \frac{1}{216} \left(\frac{L}{S}\right)^3 \frac{(EI)_T}{EI} \tag{1.4}$$

It is interesting to note that the effect on R_1 and R_2 of all the grillage parameters, namely, L, S, EI, $(EI)_T$, is reflected in a single parameter

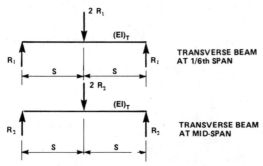

Figure 1.9 Loads on transverse beams.

η'. Since η' governs the distribution of loads, it is called the *characterizing parameter*. A treatment of the concept of characterizing parameters in bridge analysis is given in Ref. 3.

For the particular case of the grillage shown in Fig. 1.6 with a central-point load, the values of R_1 and R_2 and thence of DFs for moments and shears at sections AA and BB are calculated for a wide range of η' values. The resulting DF values for an outer girder are plotted in Fig. 1.10. A careful scrutiny of this figure will be helpful in understanding the mechanics of load distribution.

It is clear from Eq. (1.4) that larger values of η' correspond to a grillage with better load distribution characteristics. As can be seen in Fig. 1.10, the DF for both moments and shears tends to 1.0 for η' larger than about 100, suggesting that in this range all three girders share the load substantially equally although only one of them directly carries the external load. When the girders do not share the load equally, i.e., when η' is smaller than about 100, the following observations can be made about load distribution.

1. At midspan the DF for shear is always smaller than the DF for moments in the outer girder, confirming that a greater portion of shear is retained by the loaded girder than is the case for moments.

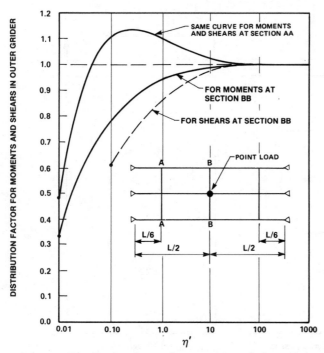

Figure 1.10 Distribution factors for moments and shears in an outer girder, at $L/6$ and $L/2$.

2. The values of DF at one-sixth span for the outer girder are always larger than the corresponding values at midspan. This indicates that the load gets distributed more evenly as one moves away from the load position. The very large differences in DF at the $L/6$ and $L/2$ longitudinal positions confirm beyond doubt that when there are two or more transverse members the value of DF does not remain even approximately constant along the span.

3. It is noted that the DF for both moments and shears for the outer girder at section AA is larger than 1.0 for a very large range of values of η'. This leads to the apparently surprising conclusion that near the support the girder carrying the direct load may often carry smaller moments and shears than the adjacent girders which do not carry an external load. The conclusion, however, is fully defensible and is supported by analyses incorporating more realistic idealizations.

It is also instructive to examine the variation along the span of distribution factors for moment and shear. For the same loading case as shown in Fig. 1.6 and with a single central-point load, i.e., with $P_1 = 0$ and $P_2 = P$, and choosing one particular value of η', say, $\eta' = 0.10$, the graphs of Fig. 1.11 are readily obtained. These graphs show

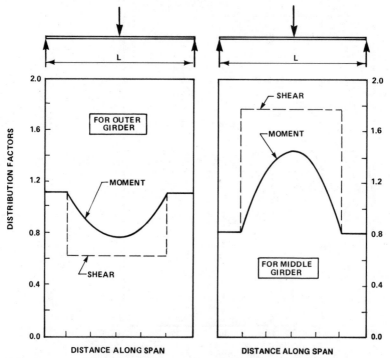

Figure 1.11 Distribution factors for moments and shears in girders of Fig. 1.4 for $\eta' = 0.10$.

that the departures from the constant value 1.00 of the DF for bending moment are accompanied by much larger departures in the DF for shear force. In Ref. 3, it has been shown that even very small variations along the span in the individual values of the DF for bending moment are accompanied by fairly large variations in those for shear.

By using DFs as the basis, a clear distinction can now be made between simplified and rigorous methods of bridge analysis. A simplified method is one which is based on the assumption that the value of DF for a girder or a longitudinal component remains constant all along the length of the bridge. Methods that fall into this category are the American Association of State Highway and Transportation Officials (AASHTO) [1] and Ontario [17] methods and others given in Refs. 3, 7, 9, and 16.

A rigorous method, on the other hand, is one which is capable of accounting for the fact that different responses are distributed in different manners. Examples of rigorous methods are dealt with in Sec. 1.2.

A claim that all rigorous methods are necessarily computer-based can be proved wrong by citing the exception of the semicontinuum method, a semigraphical manual version of which is given in Chap. 10. By using this method with relatively *simple* manual calculations, one can accurately determine the distribution of deflections, moments, and shears anywhere along the span.

For a single concentrated load, the value of DF may vary significantly along the span. However, when there are a number of loads along the span, as is the case in real-life situations, the value of DF for a given response varies between much narrower limits. In that event, the DF for longitudinal bending moment can, for all practical purposes, be regarded as constant. This is the reason why simplified methods of analysis can give results with a reasonable degree of accuracy. It should be noted, however, that the same cannot be said for DFs for other responses, such as shear force. Most simplified methods cannot directly account for differences in DF values for various responses.

1.2 Idealization of Superstructures

1.2.1 Introduction

In the analysis of a structure for load effects, such as stresses and bending moments, it is usually necessary to model the structure and the applied loads. When the analysis is done mathematically, the model of the structure is referred to as a mathematical model. When the analysis is done experimentally, the model of the structure is a physical one.

The representation of a structure or a load by a model for the purposes of mathematical analysis or model testing is referred to as *idealization*. This section deals with the mathematical idealization of bridge structures.

For many purposes, such as finding bending moments, an engineer instinctively idealizes a steel I beam as a one-dimensional beam, i.e., as a beam whose cross-sectional dimensions are ignored. The idealized beam has little similarity to the actual beam and indeed is physically impossible to construct; nevertheless, the idealization is, in many cases, a valid simplification. The actual beam, as shown in Fig. 1.12, rests on supports of finite width with nonuniform bearing pressure. The knife-edge supports which are assumed for the idealized beam and its well-defined span are therefore not exact representations of actual conditions. In spite of the idealized beam being so different from the actual one, we know by experience that this simple model is good enough for obtaining deflections, slopes, moments, and shears with a high degree of accuracy.

The mathematical model can, of course, be made to be a much closer representation of the actual structure, but this necessarily leads to increasing complexity of calculations. Such increased complexity may be warranted only if *micro* (i.e., locally detailed) distributions of the various responses are required. For example, the one-dimensional-beam model of a steel I beam cannot provide information regarding the uneven distribution of longitudinal stresses across a flange width. It will be appreciated that simple models necessarily yield responses corresponding to the macro state.

Bridge superstructures may vary widely from one structural type to another. The shallow-superstructure group includes the solid-slab, voided-slab, and slab-on-girder bridges, the cross sections of which are shown in Fig. 1.13. The cellular group includes single-cell and multicell box girders and multispine box girders, the typical cross sections of which are shown in Fig. 1.14.

When a bridge is subjected to loads which do not occupy its full width

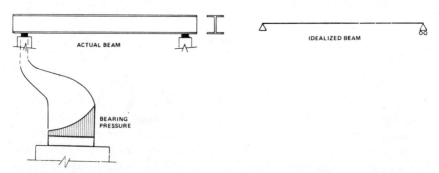

Figure 1.12 Representation of a steel beam by a one-dimensional beam.

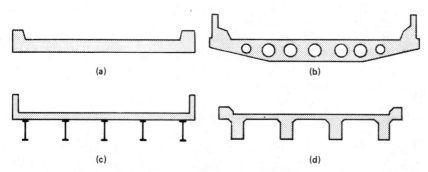

Figure 1.13 Cross section of bridge superstructures identified as shallow superstructures. (a) Slab bridge. (b) Voided-slab bridge. (c) Slab-on-girder bridge with steel girders. (d) Slab-on-girder bridge with T-beam type of construction.

(for example, when it is loaded by a vehicle), then clearly the portions of the bridge cross section immediately under the load will usually sustain a larger fraction of the load than the more remote portions. As discussed in Sec. 1.1, it can be readily appreciated that in order to account for transverse load distribution the mathematical model of the bridge cannot be a simple one-dimensional beam; such a *beam analogy* of the bridge is capable of giving the total bending moment accepted at a given cross section but cannot, by its nature, say anything about how this total is distributed. Various models, all of which are capable of accounting for the finite bridge width, are introduced in this chapter.

It is instinctively clear from an inspection of the types of bridges shown in Figs. 1.13 and 1.14 that their load distribution properties can be expected to vary quite markedly from one type to another. For example, the multicell type of cross section can be expected to be torsionally strong because of its closed boxes, while, on the other hand,

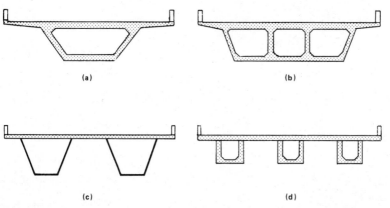

Figure 1.14 Cross sections of cellular group of bridges. (a) Single-cell bridge. (b) Multicell bridge. (c) Multispine bridge with steel boxes. (d) Multispine bridge with concrete boxes.

a slab-on-girder bridge with steel I beams can be expected to be torsionally weak. The load distribution of the former structure will be
more uniform across the bridge width than that of the latter if both
have the same spans, widths, and flexural rigidities.

Two right bridges with the same width, span, and total flexural
rigidity have cross sections as shown in Fig. 1.15. Let bridge a have
three girders each of closed cross section, and let bridge b have three
I girders. Further, let it be assumed that the outer girder of each bridge
is subjected to a load P. Ignoring small differences due to the Poisson's-
ratio effect of two-way curvature, the total deflection of three girders
at a given cross section is the same in the two bridges because they
have the same total flexural rigidity, i.e.,

$$w_{a1} + w_{a2} + w_{a3} = w_{b1} + w_{b2} + w_{b3} \tag{1.5}$$

where the notation is as shown in Fig. 1.15. The nonuniformity of
girder deflections, which is a qualitative indicator of transverse load
distribution, gives rise to girder rotations. The higher torsional rigidity
of the girders of bridge a will try to restrain these rotations, thereby
forcing the girder deflections to become more nearly uniform than is
the case in bridge b. Thus

$$\frac{w_{a1}}{w_{a2}} < \frac{w_{b1}}{w_{b2}} \tag{1.6}$$

It can be concluded that the bridge with the higher torsional rigidity
will lead to more uniform load distribution and that the bridge idealization should take account of this torsional rigidity effect.

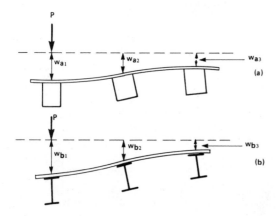

Figure 1.15 Effect of torsional rigidity of girders on load distribution. (a) Multispine bridge. (b)
Slab-on-girder bridge.

Given the various differences in load distribution characteristics of different types of bridge superstructures, several approaches have been made over the years to the question of idealization of these structures.

1.2.2   Grillage with one torsionless transverse beam

In precomputer days, a slab-on-girder bridge has been idealized as an assembly of several longitudinal beams and one torsionless transverse beam placed along the midspan of the bridge [14]. Each longitudinal beam, being an idealized one-dimensional beam, is placed along the centerline of a girder and given the flexural properties of the girder which it represents. The transverse beam, also an idealized one-dimensional beam, is assumed to be torsionless and given the total flexural rigidity of the middle half portion of the deck slab as identified in Fig. 1.16.

It is shown in Ref. 3 that in timber bridges this idealization can give moments to within about 7 percent of those obtained by a more rigorous idealization. It is recalled that the shear modulus of timber is very small as compared with its longitudinal modulus of elasticity. Because of its low shear modulus, the timber decking has very little torsional rigidity, and ignoring this factor affects the results only slightly. Concrete deck slabs, by contrast, have substantial torsional rigidities which it may not be prudent to ignore. The coarseness of the idealization of Fig. 1.16 is not suitable for accurate assessment of longitudinal shears. It is recalled that the term *longitudinal shear* refers to the vertical shear in a longitudinal strip of the bridge. In simply supported bridges, longitudinal shear is that response which at the support can be regarded as the support reaction.

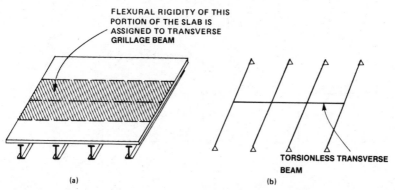

Figure 1.16   Idealization by several longitudinal beams and one torsionless transverse beam. (*a*) Actual structure. (*b*) Idealized structure.

1.2.3 Grillage analogy

The term *grillage analogy* is used for an assembly of one-dimensional beams which is subjected to loads acting in the direction perpendicular to the plane of the assembly. Unlike a plane frame, this assembly of beams incorporates torsional rigidities of the beams.

The grillage idealization is similar to the one discussed in Sec. 1.2.2, with the exception that there is no restriction on the number of transverse beams or on their torsional rigidities. The grillage analogy is applied with the help of computer programs, usually commercially available ones.

Slab bridges. As shown in Ref. 20, the deformation characteristics of a rectangular element of an isotropic plate subjected to out-of-plane loads can be represented by an assembly of six beams. Four beams lie along the edges of the element and the other two along its diagonals. This idealization is shown in Fig. 1.17. Expressions for the properties of the various beams are as follows.

$$I_x = \left(L_y - \frac{\nu L_x^2}{L_y} \right) \frac{t^3}{24(1 - \nu^2)}$$

$$I_y = \left(L_x - \frac{\nu L_y^2}{L_x} \right) \frac{t^3}{24(1 - \nu^2)}$$

$$J_x = \left[\frac{EL_y(1 - 3\nu)}{G} \right] \frac{t^3}{24(1 - \nu^2)} \qquad (1.7)$$

$$J_y = \left[\frac{EL_x(1 - 3\nu)}{G} \right] \frac{t^3}{24(1 - \nu^2)}$$

$$I_d = \left[\frac{\nu(L_x^2 + L_y^2)^{1.5}}{L_x L_y} \right] \frac{t^3}{24(1 - \nu^2)}$$

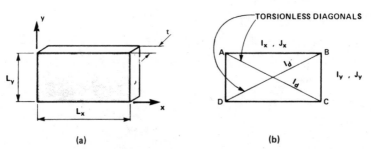

Figure 1.17 Grillage idealization of a slab element. (*a*) Slab element. (*b*) Plan of equivalent assembly of beams.

where I and J refer to the second moment of area and torsional inertia respectively. Subscripts x, y, and d refer respectively to beams along the x and y directions and to diagonal beams, v is the Poisson's ratio of the material of the plate, and other notation is as shown in Fig. 1.17.

For practical purposes, a concrete slab bridge can be regarded as an isotropic plate. Therefore, such a bridge can be conceptually divided into a number of rectangular segments, each of which can be idealized as an assembly of six beams. This idealization, as shown in Fig. 1.18a and b, can be applied to both right and skew bridges. However, as has been pointed out in Ref. 12, the resulting grillage is a complex one, the results of analysis of which are difficult to interpret. By making the Poisson's ratio zero, the diagonal beams can be eliminated, thus reducing the grillage to an orthogonal assembly of beams. Expressions for the various beam properties, corresponding to zero Poisson's ratio,

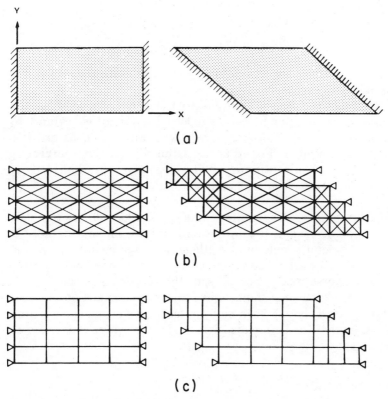

Figure 1.18 Slab idealization by grillages. (*a*) Actual slabs. (*b*) Grillage idealizations for a finite value of v. (*c*) Grillage idealizations for $v = 0$.

are as follows:

$$I_x = \frac{L_y t^3}{24}$$

$$I_y = \frac{L_x t^3}{24}$$

$$J_x = \frac{E}{G} \frac{L_y t^3}{24} \tag{1.8}$$

$$J_y = \frac{E}{G} \frac{L_x t^3}{24}$$

The effect of a nonzero value of the Poisson's ratio can be approximated by the following relationships, which, strictly speaking, are applicable only when the plate deflections are themselves independent of Poisson's ratio.

$$(M_x)_\nu = (M_x)_0 + \nu (M_y)_0$$
$$(M_y)_\nu = (M_y)_0 + \nu (M_x)_0 \tag{1.9}$$

where the subscripts ν and 0 refer to responses for finite and zero values of Poisson's ratio, respectively.

It is noted that the plate deflections are independent of the Poisson's ratio when a rectangular plate is supported along all edges and, by inference, when the slab bridge is very wide and the loads are well away from the free edges. For other cases, the guidance on interpreting the results of the semicontinuum analysis, as given in Chap. 10, can also be applied to the results of grillage analogy.

Slab-on-girder bridges. The grillage analogy works well for the slab-on-girder type of bridge. It is almost instinctive for an engineer to idealize this type of bridge by a grillage. Grillage beams in the longitudinal directions are made to coincide with the centerlines of actual girders. Transverse members represent the properties of the decking and (if present) transverse diaphragms. It is recommended that in most bridges the decking be represented by at least seven transverse beams. Recommendations on the idealization of various bridge types as grillages are given in Ref. 12.

1.2.4   Orthotropic plate

An *orthotropic plate* is a plate having constant thickness but different flexural and torsional properties in two mutually perpendicular directions. A familiar example of an actual orthotropic plate is a plywood panel in which the grains of adjacent laminates of wood run in perpendicular directions to each other. The standard notation for the various plate rigidities in a rectangular orthotropic plate is as follows:

D_x Longitudinal flexural rigidity per unit width

D_y Transverse flexural rigidity per unit length

D_{xy} Longitudinal torsional rigidity per unit width

D_{yx} Transverse torsional rigidity per unit length

D_1 Longitudinal coupling rigidity per unit width

D_2 Transverse coupling rigidity per unit length

Length is measured along the x direction, which is thus the longitudinal direction of the bridge, as shown in Fig. 1.19. Similarly, the y direction is the transverse direction, so that the width is measured along y.

The deflection of an orthotropic plate is governed by the following partial differential equation:

$$D_x \frac{\partial^4 w}{\partial x^4} + (D_{xy} + D_{yx} + D_1 + D_2) \frac{\partial^4 w}{\partial x^2 \, \partial y^2} + D_y \frac{\partial^4 w}{\partial y^4} = p(x, y) \quad (1.10)$$

where $p(x, y)$ is the intensity of load at the position x, y. It is interesting to note that the plate deflections are affected not by the individual values of the longitudinal and transverse torsional rigidities but by their sum.

Responses such as moments and shears are calculated per unit width or per unit length as the case may be. The standard notation for the

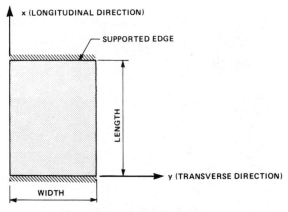

Figure 1.19 Plan of an orthotropic plate.

various plate responses is as follows:

M_x Longitudinal bending moment per unit width

M_y Transverse bending moment per unit length

M_{xy} Longitudinal torsional moment per unit width

M_{yx} Transverse torsional moment per unit length

Q_x Longitudinal shear per unit width

Q_y Transverse shear per unit length

Q_x and Q_y are also referred to as Kirchhoff's-edge reactions. Another shear force, called the supplemented shear force, is also used in orthotropic plate theory; this results from the replacement of twisting moments at the boundaries by distributions of vertical forces in order to reduce the number of boundary conditions. The two supplemented shear intensities are:

V_x Longitudinal supplemented shear per unit width

V_y Transverse supplemented shear per unit length

In standard textbooks (e.g., Refs. 7, 11, and 19), it is shown that the various plate responses are given by the following equations.

$$M_x = -\left(D_x \frac{\partial^2 w}{\partial x^2} + D_1 \frac{\partial^2 w}{\partial y^2}\right)$$

$$M_y = -\left(D_y \frac{\partial^2 w}{\partial y^2} + D_2 \frac{\partial^2 w}{\partial x^2}\right)$$

$$M_{xy} = D_{xy} \frac{\partial^2 w}{\partial x\, \partial y}$$

$$M_{yx} = -D_{yx} \frac{\partial^2 w}{\partial x\, \partial y}$$

$$Q_x = -\left[D_x \frac{\partial^3 w}{\partial x^3} + (D_{yx} + D_1) \frac{\partial^3 w}{\partial x\, \partial y^2}\right]$$

$$Q_y = -\left[D_y \frac{\partial^3 w}{\partial y^3} + (D_{xy} + D_2) \frac{\partial^3 w}{\partial y\, \partial x^2}\right]$$

$$V_x = -\left[D_x \frac{\partial^3 w}{\partial x^3} + (D_{xy} + D_{yx} + D_1) \frac{\partial^3 w}{\partial x\, \partial y^2}\right]$$

$$V_y = -\left[D_y \frac{\partial^3 w}{\partial y^3} + (D_{xy} + D_{yx} + D_2) \frac{\partial^3 w}{\partial y\, \partial x^2}\right]$$

(1.11)

Slab bridges. The isotropic plate is a specific case of the orthotropic plate in which the longitudinal rigidities are equal to their respective counterparts in the transverse direction. The idealization of a solid-slab bridge by an isotropic plate directly appeals to engineering judgment because of obvious similarities in the prototype and the model.

Slab-on-girder bridges. If one examines the orthotropic plate idealization of a slab-on-girder bridge, the strengths and weaknesses of the idealization readily emerge. For example, such a bridge is stiffer in longitudinal bending than it is in transverse bending; this is because the longitudinal girders make a large contribution to longitudinal bending stiffness whereas only the deck-slab and transverse diaphragms, if present, are effective in transverse bending. This difference in flexural rigidities can be represented in the orthotropic plate by making D_x larger than D_y. Again, the different torsional stiffnesses can be represented by appropriate choices of D_{xy} and D_{yx}. However, the slab-on-girder bridge has concentrations of longitudinal bending and torsional stiffness at the positions of the girders in the manner shown schematically in Fig. 1.20. The orthotropic plate idealization requires these large local variations to be ignored and to be replaced by stiffnesses D_x and D_{xy}, which are uniformly distributed across the bridge width.

The failure of the orthotropic plate idealization to represent fully the physical nature of the bridge has two main consequences:

1. In bridges with few girders, say, three, the bending moments and shears obtained are subject to significant errors, especially if the bridge is wide and the load occupies only a fraction of the width.

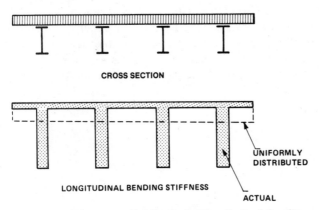

Figure 1.20 Transverse distribution of longitudinal bending stiffness in a noncomposite slab-on-girder bridge.

2. Transverse moments, which are complex combinations of bending between girders and bending due to nonuniform girder deflections, cannot be accurately obtained.

Articulated plates. A particular case of an orthotropic plate, in which D_y is equal to zero, is called the *articulated plate*. This idealization has been successfully used to analyze multibeam bridges [18] and also multispine bridges [4].

The validity of the articulated plate idealization for multispine bridges has been established in Ref. 4, where the distribution factors for longitudinal moments as obtained by the articulated plate theory and the finite strip method are compared. The latter, as discussed in Sec. 1.2.6, represents the structure very closely. It is noted that the finite strip method provides results in the form of stresses, which are then integrated to obtain longitudinal moments per spine. The moment per spine is divided by the spine spacing to obtain the intensity of longitudinal moments. It is important to note that, in spite of the close agreement between the macro responses obtained by the two methods, the articulated plate theory, like all other methods involving two-dimensional idealizations, is unable to pick up the actual variation of flange stresses which are illustrated in Fig. 1.21.

1.2.5 Finite elements

As commonly used in bridge analysis, a finite element can be a one-dimensional beam, or a segment of a plate subjected to pure bending, or a segment of a plane shell subjected to both bending and planar forces. These elements are shown in Fig. 1.22.

The finite element method consists of solving the mathematical model which is obtained by idealizing a structure as an assembly of the various elements. The grillage analogy method given in Sec. 1.2.3 is a particular case of the finite element method in which the bridge superstructure is idealized as an assembly of one-dimensional-beam elements. Quite often, however, the use of the finite element method implies idealization by continuum elements, such as the plate or shell elements discussed above.

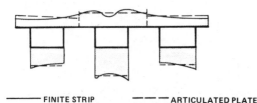

————— FINITE STRIP   — — —ARTICULATED PLATE

Figure 1.21 Longitudinal flange stresses.

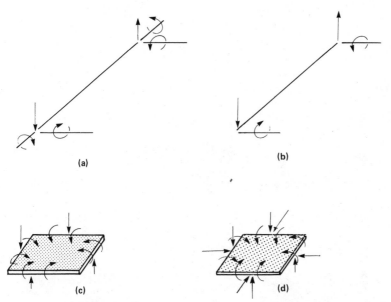

Figure 1.22 Examples of finite elements. (a) Discrete beam element for grillage. (b) Discrete beam element for plane frame. (c) Plate bending element. (d) Plane shell element.

For analysis by the finite element method incorporating continuum elements, a bridge superstructure can be idealized either by plate bending elements or by plane shell elements. For the former, the structure is essentially idealized as an orthotropic plate with the same plate rigidities as given in Sec. 1.2.4. The idealization by plane shell elements can be three-dimensional and, therefore, closer to reality in certain cases. The idealization of a segment of a box-girder bridge by plane shell elements is shown in Fig. 1.23.

Following the development of the finite element method over the past three decades, there now exists an extensive literature on it. Reference 13 will be found useful in the selection of the various elements for bridge analysis. References 8, 21, and 22 also deal with the relevance of the finite element method to bridge superstructures.

The power and wide applicability of the finite element method are well known and require no further elaboration here. It should be noted, however, that the properties of individual elements must be chosen with great care if the results of the analysis are to represent closely the actual behavior of the bridge; in particular, one should guard against the assumption that increasing the fineness of the mesh of the elements will always and unfailingly result in convergence to the actual behavior of the bridge.

There is no doubt that in skilled hands the finite element method is

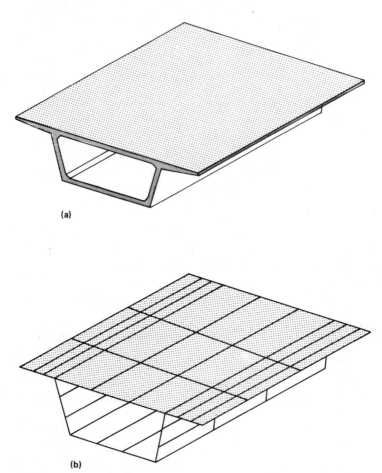

(a)

(b)

Figure 1.23 Idealization of a box girder by plane shell elements. (*a*) Actual structure. (*b*) Idealized structure.

a preferred tool. For this reason some major bridge design authorities are developing large-scale analysis systems using the method; such analysis systems can then be made specific to the requirements of a particular code of practice. Thus, a large-computer procedure is emerging in which a bridge designer will have available the services of a central analysis bureau of some kind and in which that central bureau will use the finite element method to perform needed analysis, using a large central computer for the calculations.

The characteristics of this procedure, and indeed of the finite element method per se, as applied to bridges, are that (1) the results of the analysis are in numerical form and difficult to generalize, so that modification of a design will frequently involve partial or complete reanalysis; and (2) for reasonably close representation of the bridge a large number of elements are needed, so that the availability of a large computer is indispensably necessary.

1.2.6 Finite strips

The finite strip method is a particular case of the finite element method in which the *element* is in the form of a strip extending, in the case of a bridge, from abutment to abutment. By using this method a bridge superstructure can be idealized as a three-dimensional assembly of strips. An example of the idealization of a slab-on-girder bridge by finite strips is shown in Fig. 1.24.

Because of its significantly lower computational requirements, the finite strip method will prove to be more economical for certain structures than the general finite element method. As in the case of the finite element method, there exists an extensive published literature on the finite strip method. References 5, 6, and 15 will be found especially useful by bridge designers.

1.2.7 Semicontinuum idealization

By reviewing the physical characteristics of the bridge types shown in Figs. 1.13 and 1.14, it will be clear that in all cases the transverse bending stiffness and transverse torsional stiffness are spread, substantially uniformly, along the length of the bridge. On the other hand, with the exception of the solid-slab type, most of the bridges have significant concentrations of longitudinal bending and torsional stiffness at identifiable locations. For the voided-slab bridge these locations are the web positions defined between adjacent voids, for the slab-on-girder type they are the positions of the longitudinal girders, and for

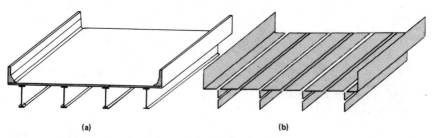

(a)   (b)

Figure 1.24 Idealization of a slab-on-girder bridge by finite strips. (*a*) Actual structure. (*b*) Idealized structure.

the cellular types they are the positions of the webs. These considerations lead to an idealization in which the longitudinal bending and torsional stiffness of the bridge are concentrated in a number of one-dimensional longitudinal beams while the transverse bending and torsional stiffness are uniformly spread along the length in the form of an infinite number of transverse beams, which thereby constitute a *transverse medium*. The idealization can thus be regarded as a special case of the grillage analogy in which the number of transverse beams approaches infinity, as shown schematically in Fig. 1.25. Because the properties of the bridge are represented discretely insofar as longitudinal stiffness is concerned and continuously insofar as transverse stiffness is concerned, this idealization is semicontinuum in nature. In an early form of the semicontinuum idealization developed some years ago, the torsional rigidities of the longitudinal beams and the deck were either ignored completely or, in the case of longitudinal effects, were handled only approximately [10]. Most of the methods of analysis presented in this book are based on a generalized form of the semi-continuum idealization which can take proper account of torsional rigidities in both the longitudinal and the transverse directions. The longitudinal girders of the idealization are usually placed at the same positions, if any, as the actual girders of the bridge so as to keep the mathematical model of the bridge as close as possible to the physical reality. Frequently, all these girders will have the same flexural stiffness (EI) and torsional rigidity (GJ). However, this is not obligatory; for example, if the bridge has significant edge stiffening, this can be represented in the mathematical model by making its edge girders correspondingly stiffer than the interior ones. In the case of a solid-slab bridge, the longitudinal girders are usually taken to be all the same; for accurate representation of the physical reality the number of such girders is preferably taken to be at least five.

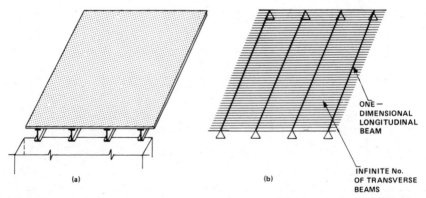

Figure 1.25 Idealization by a semicontinuum. (*a*) Actual bridge. (*b*) Semicontinuum idealization.

Adoption of the semicontinuum idealization. For the derivation of the methods of analysis presented in this book a key element is the adoption of harmonic analysis of externally applied loads, which is discussed in Chap. 2. The second key element is the adoption of the semicontinuum idealization of the bridge. As will rapidly emerge in Chap. 3, the effect of these two key features taken together is to reduce to a small fraction of their former number the number of unknowns which must be carried in a computation in order to obtain a given level of accuracy.

REFERENCES

1. American Association of State Highway and Transportation Officials: *Specifications for Highway Bridges*, Washington, 1983.
2. Bakht, B.: Statistical analysis of timber bridges, *ASCE Journal of Structural Division*, 109(8), 1983, pp. 1761–1779.
3. Bakht, B., and Jaeger, L. G.: *Bridge Analysis Simplified*, McGraw-Hill, New York, 1985.
4. Cheung, M. S., Bakht, B., and Jaeger, L. G.: Analysis of box girder bridges by grillage and orthotropic plate methods, *Canadian Journal of Civil Engineering*, 9(4), 1982, pp. 595–601.
5. Cheung, M. S., Cheung, Y. K., and Ghali, A.: Analysis of slab and girder bridges by the finite strip method, *Building Science*, 5, 1970, p. 95.
6. Cheung, Y. K.: *Finite Strip Method in Structural Analysis*, Pergamon, Oxford, 1976.
7. Cusens, A. R., and Pama, R. P.: *Bridge Deck Analysis*, Wiley, London, 1975.
8. Davies, J. D., Somerville, I. J., and Zienkiewicz, O. C.: Analysis of various types of bridges by the finite element method, *Proceedings, Conference on Developments in Bridge Design and Construction, Cardiff*, Crossby Lockwood, London, 1971.
9. Guyon, Y.: Calcul des ponts larges à poutres multiples solidarisées par des entretoises, *Annales des Ponts et Chaussées*, no. 24, 1946, pp. 553–612.
10. Hendry, A. W., and Jaeger, L. G.: *The Analysis of Grid Frameworks and Related Structures*, Prentice-Hall, Englewood Cliffs, N.J., 1958.
11. Jaeger, L. G.: *Elementary Theory of Elastic Plates*, Pergamon, Oxford, 1964.
12. Jaeger, L. G., and Bakht, B.: The grillage analogy in bridge analysis, *Canadian Journal of Civil Engineering*, 9(2), 1982, pp. 224–235.
13. Jategaonkar, R., Jaeger, L. G., and Cheung, M. S.: *Bridge Analysis Using Finite Elements*, Canadian Society for Civil Engineering, 1985.
14. Leonhardt, F., and Andre, W.: *Die vereinfachte Traperrost Berechnung*, Julius Hoffman Press, Stuttgart, 1950.
15. Loo, Y. C., and Cusens, A. R.: Development of the finite strip method in the analysis of bridge decks, *Proceedings, Conference on Developments in Bridge Design and Constructions, Cardiff*, Crossby Lockwood, London, 1971.
16. Morice, P. B., and Little, G.: *The Analysis of Right Bridge Decks Subjected to Abnormal Loading*, Report Db 11, Cement and Concrete Association, London, 1956.
17. *Ontario Highway Bridge Design Code*, Ministry of Transportation and Communications, Downsview, Ontario, 1983.
18. Pama, R. P., and Cusens, A. R.: Edge beam stiffening of multi-beam bridges, *ASCE Journal of Structural Division*, 93(ST2), 1967, pp. 141–161.
19. Timoshenko, S. P., and Woinowsky-Krieger, S.: *Theory of Plates and Shells*, McGraw-Hill, New York, 1959.
20. Yettram, A. L., and Husain, M. H. A.: Grid framework method for plates in flexure, *ASCE Journal of Engineering Division*, 91(EM3), 1965, pp. 53–64.
21. Zienkiewicz, O. C.: *The Finite Element Method in Engineering Science*, McGraw-Hill, New York, 1971.
22. Zienkiewicz, O. C., and Cheung, Y. K.: The finite element method for analysis of elastic isotropic and orthotropic slabs, *Proceedings, Institution of Civil Engineers*, August 1964.

Harmonic Analysis of Beams

As discussed in Sec. 1.2, for the purposes of analysis a bridge super-structure must usually be idealized as a mathematical model such as an assembly of finite elements, an orthotropic plate, or a semicontin-uum. Similarly, applied loads also require appropriate mathematical idealization. For example, loads concentrated over relatively small lengths are frequently idealized as point loads for the purposes of beam analysis. Such point loads, like the actual loads, are discontinuous functions with respect to the span of the beam. It is possible, and sometimes convenient for analysis, to represent a discontinuous load on a beam as a continuous function or a series of continuous functions by means of a harmonic series.

For the methods of analysis presented in this book, the wheel loads of a design vehicle are analyzed into harmonic components. In order to make effective use of these methods, the designer must become familiar with this technique, which accordingly is discussed in this chapter. Simply supported and multispan beams that are subjected to concentrated loads can be conveniently analyzed on a personal computer by the harmonic series method presented below. It is noted that, in spite of intimidating-looking equations (which the engineer is not, in any event, called upon to use directly in the analysis), the concept of harmonic analysis is simple. It can be quickly grasped by a structural engineer who understands elementary beam theory.

2.1 Point Loads on Simply Supported Beams

As shown in App. I, a point load P on a simply supported beam of span L can be represented as a continuous load of intensity given by the following expression:

$$p_x = \frac{2P}{L}\left(\sin\frac{\pi c}{L}\sin\frac{\pi x}{L} + \sin\frac{2\pi c}{L}\sin\frac{2\pi x}{L} + \cdots\right) \qquad (2.1)$$

where c is the distance of the point load from the left-hand support and x is the distance along the span, also measured from the left-hand support.

Thus according to Eq. (2.1), a point load is equivalent to the sum of an infinite number of distributed loads, each of which corresponds to a term of the series and is a continuous function of x. For example, if we substitute $c = L/4$ in Eq. (2.1), we obtain the representation of a point load at the quarter-span position. This series is diagrammatically shown in Fig. 2.1. As shown in the figure, the load corresponding to the first term (or harmonic) has the shape of a half sine wave, and that

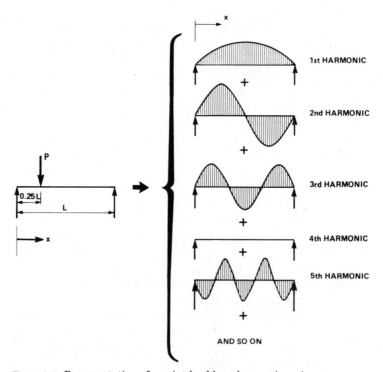

Figure 2.1 Representation of a point load by a harmonic series.

corresponding to the second has the shape of two half sine waves, and so on. For the particular case shown in Fig. 2.1, the contribution to the p_x series is zero for harmonic numbers that are divisible by 4.

Figure 2.2 shows the load corresponding to the thirtieth harmonic of the point load P at quarter span. With the help of this figure it can be readily appreciated that when n is large, a load having the shape $\sin (n\pi x/L)$ comprises a rapid alternation of positive and negative distributed loads which tend to cancel each other out. When the harmonic number is large enough, the net effect of these upward and downward loads on beam bending moments and deflections is negligible.

From elementary small-deflection beam theory it is well known that the load intensity p_x, bending moment M_x, shear force V_x, and slope θ_x of a beam of uniform flexural rigidity EI are related to its deflection w by the following equations.

$$p_x = EI \frac{d^4w}{dx^4}$$

$$V_x = -EI \frac{d^3w}{dx^3}$$

$$M_x = -EI \frac{d^2w}{dx^2}$$

(2.2)

$$\theta_x = \frac{dw}{dx}$$

Thus values of V_x, M_x, θ_x, and w can be obtained by successively integrating the right-hand side of Eq. (2.1) with respect to x. The end conditions of a simply supported beam are such that all constants of integration are equal to zero, and the following equations are obtained.

$$V_x = \frac{2P}{\pi} \sum_{n=1}^{n=\infty} \frac{1}{n} \sin \frac{n\pi c}{L} \cos \frac{n\pi x}{L}$$

$$M_x = \frac{2PL}{\pi^2} \sum_{n=1}^{n=\infty} \frac{1}{n^2} \sin \frac{n\pi c}{L} \sin \frac{n\pi x}{L}$$

(2.3)

$$\theta_x = \frac{2PL^2}{\pi^3 EI} \sum_{n=1}^{n=\infty} \frac{1}{n^3} \sin \frac{n\pi c}{L} \cos \frac{n\pi x}{L}$$

$$w = \frac{2PL^3}{\pi^4 EI} \sum_{n=1}^{n=\infty} \frac{1}{n^4} \sin \frac{n\pi c}{L} \sin \frac{n\pi x}{L}$$

where EI is the flexural rigidity of the beam.

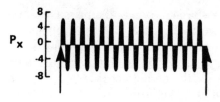

Figure 2.2 Load corresponding to the thirtieth harmonic.

Figure 2.3 shows the values of p_x, V_x, M_x, and EIw for the first harmonic in a beam of 20-unit length and subjected to a load of 1000 units at the quarter span. The figure also compares these first harmonic values with the respective true solutions. These true solutions are shown in dashed lines and are, of course, statically determinate. They

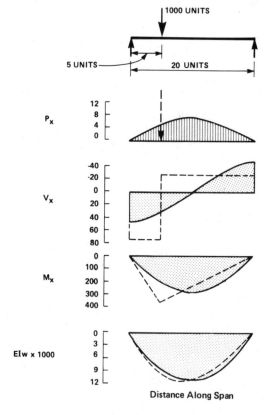

Figure 2.3 Load and various responses of a simply supported beam due to the first term of the harmonic series for a point load.

are referred to in this book as *free solutions*; for example, the free-bending-moment diagram is the familiar triangular one. The free bending moment is designated as M_L, the free shear force as V_L, etc. It can be seen in Fig. 2.3 that the first harmonic component of p_x does not resemble the concentrated load at all, yet the deflections due to this first harmonic are fairly close to the actual (free) ones.

By adding the effects of higher harmonics, it can be readily demonstrated that responses corresponding to higher derivatives of deflections converge more slowly to the true (free) solutions. For example, Fig. 2.4 gives the values of p_x, V_x, M_x, and EIw due to the second harmonic alone and also that due to the first two harmonics. The deflection due to the latter loading has now converged to almost exactly

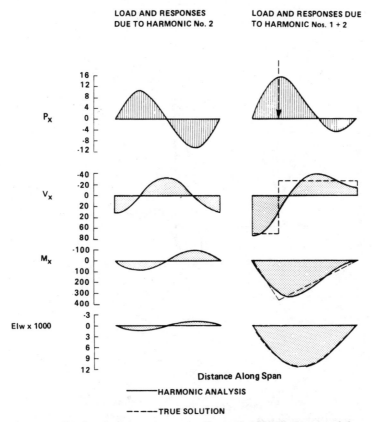

Figure 2.4 Load and various responses due to the first two terms of the harmonic series for a point load. (For details of load and beam see Fig. 2.3.)

the free shape, but the bending moment has some way to go and the shear force even further.

The effect of increasing the number of harmonics to five can be seen in Fig. 2.5. The contribution of the fifth harmonic to moments is now very small, and to the deflections it is so small that it cannot be shown in the figure. Clearly, the first five harmonics are quite sufficient insofar as deflections are concerned. It should be noted, however, that although the maximum shear obtained from harmonic analysis is quite close to the true value, the two shear diagrams are still not close enough, thus indicating that more harmonics are needed for a satisfactory representation of V_x.

V_x and M_x due to the first 30 harmonics are plotted in Fig. 2.6 and compared with the respective true responses. From the comparisons

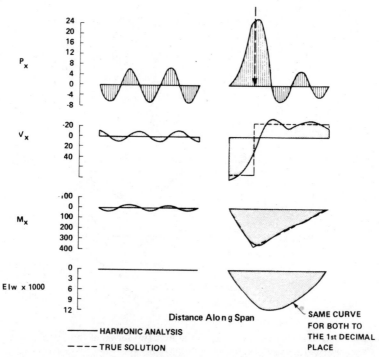

Figure 2.5 Load and various responses due to the first five terms of the harmonic series for a point load. (For details of load and beam see Fig. 2.3.)

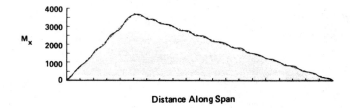

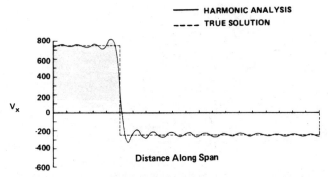

Figure 2.6 Responses due to the first 30 harmonics of a point load. (For details of load and beam see Fig. 2.3.)

shown in the figure it can be seen that the results of harmonic analysis now compare fairly well with the true values even in the case of shear.

2.2 Uniformly Distributed Loads on Simply Supported Beams

It is shown in App. I that a uniformly distributed load on a simply supported beam of span L is represented by the following expression:

$$p_x = \frac{4q}{\pi} \left(\sin \frac{\pi x}{L} + \frac{1}{3} \sin \frac{3\pi x}{L} + \frac{1}{5} \sin \frac{5\pi x}{L} + \cdots \right) \quad (2.4a)$$

That is,

$$p_x = \frac{4q}{\pi} \sum_{n=1,3,5}^{n=\infty} \frac{1}{n} \sin \frac{n\pi x}{L} \quad (2.4b)$$

where q is the intensity of the uniformly distributed load. The equations for the various beam responses can then be obtained by successively integrating the right-hand side of Eq. $(2.4a)$ or Eq. $(2.4b)$ with respect to x. As was the case for the single-point load, the various constants of integration are all found to be equal to zero, and the following equations are obtained.

$$V_x = \frac{4qL}{\pi^2} \sum_{n=1,3,5}^{n=\infty} \frac{1}{n^2} \cos \frac{n\pi x}{L}$$

$$M_x = \frac{4qL^2}{\pi^3} \sum_{n=1,3,5}^{n=\infty} \frac{1}{n^3} \sin \frac{n\pi x}{L}$$

$$\theta_x = \frac{4qL^3}{\pi^4 EI} \sum_{n=1,3,5}^{n=\infty} \frac{1}{n^4} \cos \frac{n\pi x}{L}$$

$$w = \frac{4qL^4}{\pi^5 EI} \sum_{n=1,3,5}^{n=\infty} \frac{1}{n^5} \sin \frac{n\pi x}{L}$$

(2.5)

Thus, for example,

$$M_x = \frac{4qL^2}{\pi^3} \left(\sin \frac{\pi x}{L} + \frac{1}{3^3} \sin \frac{3\pi x}{L} + \frac{1}{5^3} \sin \frac{5\pi x}{L} + \cdots \right) \quad (2.6)$$

The representation of p_x and the various beam responses due to the first term of the series are compared with the respective true values in Fig. 2.7. It can be seen that the correspondence between the true values and the values obtained from the first harmonic analysis is

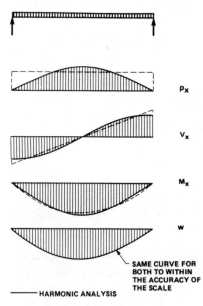

p_x

V_x

M_x

w

SAME CURVE FOR BOTH TO WITHIN THE ACCURACY OF THE SCALE

———— HARMONIC ANALYSIS

- - - - TRUE SOLUTION

Figure 2.7 Load and various responses of a simply supported beam by the first term of the harmonic series for a uniformly distributed load.

much better than was the case for a single-point load (shown in Fig. 2.3). Indeed, the two curves for deflections are so close that on the scale of Fig. 2.7 they cannot be shown apart.

As mentioned earlier, one of the reasons for presenting discrete loads (those loads which are discontinuous with respect to the span) by a harmonic series is that the mathematical idealization provides a continuous-load function with respect to the span. In the case of a uniformly distributed load covering the entire length of the span, the load is already a continuous function of the span, and it is therefore relevant to ask why such a load should be analyzed into a harmonic series. The reason is that each term of the harmonic series has the property that a load of that shape will produce a deflection of the same shape. This property is of indispensable importance in establishing the load distribution properties of the bridges, as will be seen in Chap. 3 and subsequent chapters of this book. The behavior is clearly shown in Fig. 2.4, for example, in which a second harmonic load component is seen to give rise to a deflection component of the same shape. As can be seen in the figure, not only the beam deflection but also the moments due to the second harmonic are in the shape of the two half sine waves. Similarly, for the fifth harmonic loading, which is in the shape of five half sine waves (as shown in Fig. 2.5), the beam deflections and moments are also in the shape of five half sine waves. It is noted that

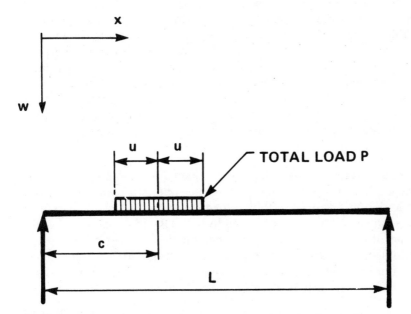

Figure 2.8 Load uniformly distributed over part of the span of a simply supported beam.

because of very small magnitudes the deflections could not be shown on the scale of the figure.

It is again readily shown, by using results given in App. I, that a load uniformly distributed over a length $2u$ (which is smaller than the span L) of a simply supported beam as shown in Fig. 2.8 can be represented by the following series expression.

$$p_x = \frac{2P}{\pi u} \sum_{n=1}^{n=\infty} \frac{1}{n} \sin \frac{n\pi c}{L} \sin \frac{n\pi u}{L} \sin \frac{n\pi x}{L} \tag{2.7}$$

where P is the total load and c is the distance of the center of gravity of the load from the same support from which x is measured.

2.3   Convergence of the Series

When loads are represented by a harmonic series, the expressions for the various beam responses are also in the form of series, with an infinite number of terms. Before undertaking analysis by such expressions, it is desirable to have an idea of the minimum number of terms which would yield results of required accuracy. It has been noted earlier in this chapter that only a few terms of the relevant series are required to obtain deflections with good accuracy. In order to obtain, say, shears with the same good degree of accuracy, a relatively large number of harmonics should be included in the calculations.

The problem of convergence of the various responses to within acceptable margins of accuracy is discussed in this section. The measure of convergence for a response is taken as the ratio of the response obtained by harmonic analysis to the respective true solution, the latter being obtained by elementary beam theory. Δ_w, Δ_M, and Δ_V are taken as the measures of convergence for deflections, moments, and shears respectively. Clearly, when the response obtained by the harmonic analysis and the true solution are the same, the measure Δ is equal to 1.0.

Single-point loads versus uniformly distributed loads. Δ_V, Δ_M, and Δ_w, corresponding to the maximum responses due to a point load at midspan and a uniformly distributed load covering the entire span of the simply supported beam, are plotted in Fig. 2.9 with respect to the total number of harmonics included in the calculations. Strictly speaking, the plots should have shown only points corresponding to the various harmonic numbers. These points are joined by straight lines to facilitate reading. Several observations can be made:

1. For a given loading, responses corresponding to lower derivatives of deflection with respect to x converge faster than those corresponding to higher derivatives.

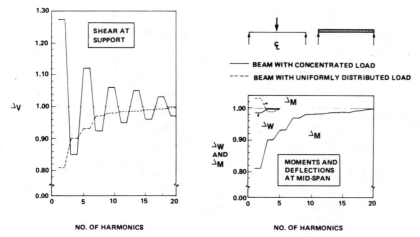

Figure 2.9 Convergence of maximum beam responses under single concentrated and uniformly distributed loads.

2. A response due to the distributed load converges faster than the same response due to a concentrated load.

3. The rate of convergence of moments due to a concentrated load is the same as that for shears due to the distributed load.

A careful scrutiny of Eqs. (2.3) and (2.5) will show that the above observations are indeed predictable.

The convergence of a series depends upon the magnitude of its successive terms. The smaller the ratio between successive terms, the faster the series converges. It can be seen from Eq. (2.3) that for the single-point-load case the successive terms of the series for shears correspond to the inverse of the harmonic number while those for deflections correspond to the inverse of the fourth power of the harmonic number. Consequently, the ratios of successive terms of the deflection series are much smaller than those for the shear series, thus causing the deflections to converge faster.

A comparison of corresponding equations for the point-load case and the distributed-load case [Eqs. (2.3) and (2.5)] will show that in the latter the denominator has a higher power of n. For example, the moment equation for a point load has n^2 in the denominator, while that for a distributed load has n^3. This explains why the same response converges faster in the latter case. Also, it will be noted that the expression for shears due to the distributed loads has n^2 in the denominator [Eq. (2.5)], as does the expression for moments due to concentrated load [Eq. (2.3)], explaining why moments under a concentrated load and shears under a uniformly distributed load converge in a similar manner.

Multiple-point loads.   It has been established that responses due to a uniformly distributed load converge faster than those due to a single-point load. Therefore, it can be confidently foreseen that convergence should improve with an increase in the number of loads, which in the limiting case of an infinite number of loads are equal to distributed loads. However, a general quantitative assessment of the rate of convergence is difficult without knowing the number and position of loads, the length of the span, and the position of the reference point.

Δ_V and Δ_M are plotted against the number of harmonics in Figs. 2.10 and 2.11 for two simply supported beams under three-point loads which correspond to a bridge design vehicle. One beam has a span of 30 m (98 ft) and the other a span of 100 m (328 ft). Δ_V and Δ_M in Fig. 2.10 correspond to the left-hand support and midspan respectively; in Fig. 2.11 both Δ_V and Δ_M correspond to the quarter span.

It can be readily concluded from the diagrams given in Figs. 2.10 and 2.11 that the convergence of results is faster in the shorter beam, thus confirming that it is not just the number of point loads that governs convergence. The convergence is also affected by the spacings of these loads with respect to the beam span.

2.4   Multispan Beams

The harmonic series methods described in Secs. 2.1 and 2.2 can also be used to analyze a multispan beam as described below.

In the case of a multispan beam with simply supported end supports, the redundancy is caused by the reactions of the intermediate supports. If these reactions can somehow be determined, the statically indeter-

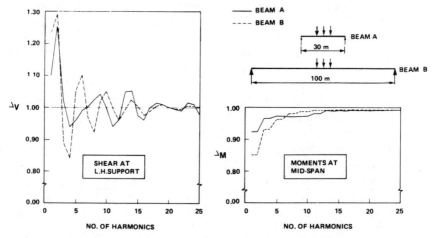

Figure 2.10   Convergence of shears at support and moments at midspan.

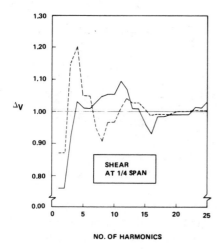

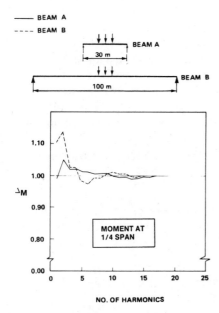

Figure 2.11 Convergence of shears and moments at one-fourth span.

minate problem of the multispan beam is reduced to the statically determinate one of a simply supported beam subjected to externally applied downward loading and (usually upward) intermediate-support reactions. This process, which is known as the *force method,* is schematically shown in Fig. 2.12. As shown in the figure, the span of the above-mentioned simply supported beam is the distance between the end supports.

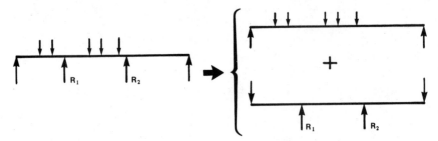

Figure 2.12   Force method of solving the problem of multispan beams.

The first step in determining the redundant-support reactions is to remove the intermediate supports and calculate deflections at the support locations due to the applied loading. This is easily done by the relevant method of those described in Secs. 2.1 and 2.2. The intermediate-support reactions which will bring the load deflections at the various support locations back to zero or to a defined settlement value are the required reactions. These are calculated as follows.

Consider the general case of a beam with $(m + 1)$ spans. The intermediate supports of such a beam are numbered from 1 to m as shown in Fig. 2.13. The intermediate supports have prescribed settlements $\delta_1, \delta_2, \ldots, \delta_m$, and the flexibilities of the supports are denoted by f_1, $f_2, \ldots, f_m$ and the support reactions by $R_1, R_2, \ldots, R_m$. The subscripts to δ, f, and R refer to the support numbers. The total vertical deflection at a support i is equal to $(\delta_i + R_i f_i)$.

The applied loading deflections of the beam without the intermediate

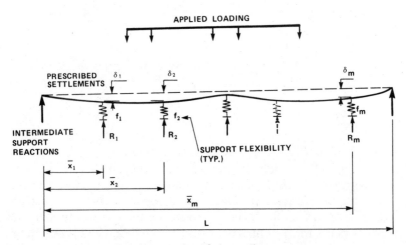

Figure 2.13   Notation for a beam unit with $(m + 1)$ spans.

support at the various support locations are denoted as $\overline{w}_1$, $\overline{w}_2, \ldots, \overline{w}_m$. The subscripts again refer to the support numbers.

The simply supported beam deflection at support i due to a unit reaction at support j is denoted by w_{ij}. The net deflections due to applied loading and support reactions at support i must be equal to $R_i f_i + \delta_i$, that is,

$$\overline{w}_i - (R_1 w_{i1} + R_2 w_{i2} + \cdots + R_m w_{im}) = R_i f_i + \delta_i \qquad (2.8)$$

The resulting set of equations for the multispan beam can be written in the following matrix form:

$$\begin{bmatrix} w_{11} + f_1 & w_{12} & \cdot & w_{1m} \\ w_{21} & w_{22} + f_2 & \cdot & w_{2m} \\ \vdots & \vdots & \vdots & \vdots \\ w_{m1} & w_{m2} & \cdot & w_{mm} + f_m \end{bmatrix} \begin{Bmatrix} R_1 \\ R_2 \\ \vdots \\ R_m \end{Bmatrix} = \begin{Bmatrix} \overline{w}_1 - \delta_1 \\ \overline{w}_2 - \delta_2 \\ \vdots \\ \overline{w}_m - \delta_m \end{Bmatrix}$$

$$(2.9)$$

or $\qquad\qquad [A]\{R\} = \{D\} \qquad\qquad (2.10)$

The various deflections for the simply supported beam can be obtained by Eq. (2.1), following which the redundant-support reactions can be obtained by solving the set of equations identified above. Once the support reactions are known, the problem of the multispan beam is reduced to that of a simply supported beam subjected to the applied loading and the support reactions.

It should be noted that the above method may be less efficient than some of the other well-known methods of continuous-beam analysis. It is given as a ready means of introducing the reader to the principles which are used in Chap. 5 for bridges with random intermediate supports.

Convergence of reactions. As can be expected, the number of harmonics required for obtaining the intermediate-support reactions to within a given degree of accuracy increases with the number of intermediate supports. The measure of convergence Δ_R for intermediate-support reactions, which is the ratio of the true and calculated support reactions, is plotted in Fig. 2.14 against the number of harmonics for beams with various numbers of spans corresponding to a single-point load. It can be seen that the reactions for the first three harmonics in beams with three or more spans are not even close to the actual reactions. However, the situation changes rapidly with the increase in the number of harmonics.

An idea of the actual number of harmonics required for the various cases can be obtained from Fig. 2.15, which is based on the convergence criterion that the intermediate-support reactions are assumed to have converged if their values do not vary by more than 0.1 percent for three successive harmonics. Figure 2.15 was constructed from the results of analyses of a number of beams having from two to five spans and subjected to one and five concentrated loads. The minimum number of harmonics required to obtain the intermediate-column reactions, according to the convergence criterion given above, is plotted against the number of intermediate supports.

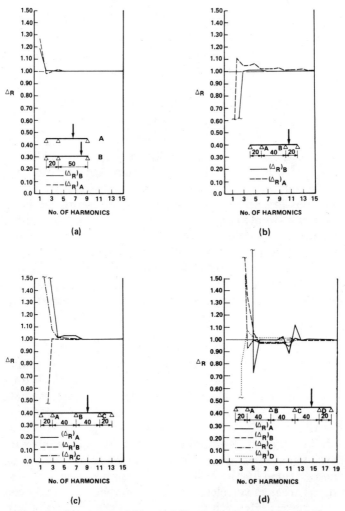

Figure 2.14   Convergence of intermediate-support reactions. (*a*) Two-span beams. (*b*) Three-span beams. (*c*) Four-span beams. (*d*) Five-span beams.

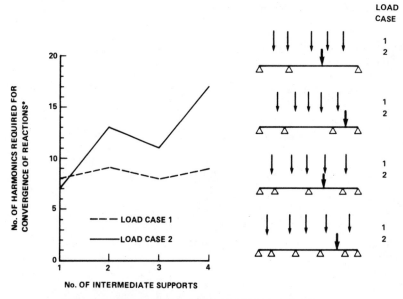

Figure 2.15 Number of harmonics versus number of intermediate supports for a certain degree of convergence.

It can be seen that the number of harmonics generally decreases as the number of loads increases. Judging from the results plotted in Fig. 2.15, it can be confidently recommended that 10 harmonics are sufficient for obtaining acceptably accurate intermediate-support reactions in multispan beams of up to five spans subjected to a series of point loads.

The Semicontinuum Method for Torsionless Bridges

As discussed in Sec. 1.2.7, the semicontinuum idealization has been used in the past for bridges with negligible torsional rigidities in both longitudinal and transverse directions. Details of this idealization and the method of analysis are given in Ref. 2. This chapter develops the *torsionless method* in some detail. Subsequent chapters of the book are based upon a generalized version of the method which utilizes the semicontinuum idealization for bridges in conjunction with nonzero values of torsional rigidities in both longitudinal and transverse directions.

The method for torsionless bridges, given in this chapter, provides an introduction and background for the generalized method. Engineers interested in developing a step-by-step understanding of the general method are encouraged to read this chapter; others may safely proceed directly to the generalized method, which, of course, includes torsionless behavior as one particular case.

3.1 First Harmonic Behavior in a Three-Girder Bridge

Line load on the center girder. We consider, as an introductory example, the semicontinuum idealization for a three-girder bridge. In this simple example, the three girders, as shown in Fig. 3.1, are taken to be all the same, with flexural stiffness EI. Thus, the total longitudinal flexural stiffness of the entire bridge is divided into three equal parts, and one part is allocated to each girder. The girders are numbered from left to right. If the total transverse flexural stiffness of the bridge is called $(EI)_T$, then this total is distributed uniformly along the length of the bridge. In an element of length δx the transverse flexural stiffness is therefore $[(EI)_T \, \delta x]/L$, where L is the span of the bridge.

The three-girder idealization is subjected to an external line load

$P \sin(\pi x/L)$ on the middle girder as shown in Fig. 3.1. Since the girders and transverse beams are taken to be torsionless, the interactions between the girders and the transverse medium consist simply of vertical forces. It is readily verified that all the conditions of equilibrium and deflection compatibility may be satisfied by taking the interactive line loading between girder 1 and the transverse medium to be $P_1 \sin(\pi x/L)$ as shown in Fig. 3.2. By symmetry the same loading appears on girder 3. Since girder 1 is simply supported, it is readily verified that at a distance x from the left-hand end it sustains a bending moment M_1 given by

$$M_1 = -EI\frac{d^2 w_1}{dx^2} = \frac{P_1 L^2}{\pi^2}\sin\frac{\pi x}{L} \tag{3.1}$$

where w_1 is the deflection of girder 1 and, by symmetry, also that of girder 3. Hence girders 1 and 3 both undergo deflections, obtained by integrating Eq. (3.1) twice:

$$w_1 = a_1 \sin\frac{\pi x}{L} \tag{3.2}$$

where
$$a_1 = \frac{P_1 L^4}{\pi^4 EI} = \frac{P_1}{k} \tag{3.3}$$

with $\pi^4 EI/L^4$ denoted by k.

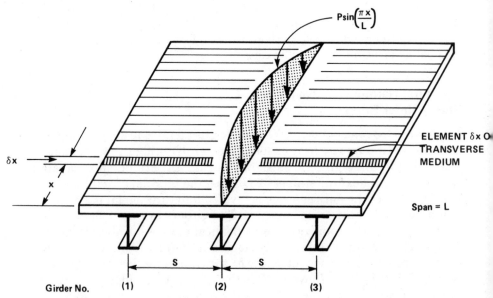

Figure 3.1 A three-girder bridge.

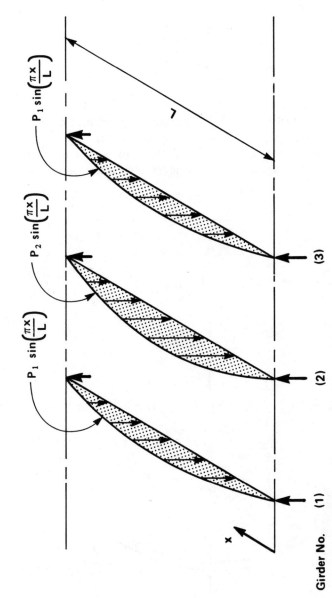

Girder No. (1) (2) (3)

Figure 3.2 Loads accepted by three girders.

Similarly, if the net downward loading on girder 2 is $P_2 \sin (\pi x/L)$, then this girder assumes a deflected shape:

$$w_2 = a_2 \sin \frac{\pi x}{L} \qquad (3.4)$$

where

$$a_2 = \frac{P_2}{k} \qquad (3.5)$$

In order to obtain P_1 and P_2, the equilibrium and deflection of the portion of the transverse medium shown in Fig. 3.1 are considered. Figure 3.3a shows the forces and deflections concerned. The flexural rigidity of this element of the transverse medium, as noted above, is

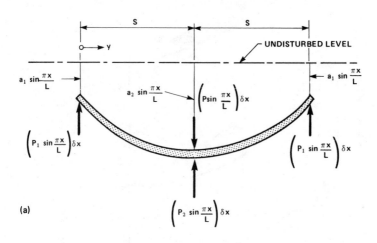

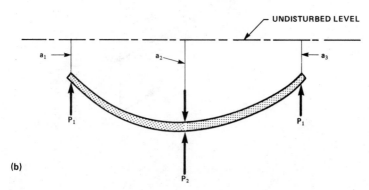

Figure 3.3 Bending of transverse elements. (a) Element of flexural rigidity $[(EI)_T/L] \, \delta x$. (b) Element of flexural rigidity D_y.

$[(EI)_T \, \delta x]/L$. Resolving vertically for equilibrium, there results

$$\left(2P_1 \sin \frac{\pi x}{L} \right) \delta x + \left(P_2 \sin \frac{\pi x}{L} \right) \delta x = \left(P \sin \frac{\pi x}{L} \right) \delta x \qquad (3.6)$$

A common factor cancels through Eq. (3.6), giving

$$2P_1 + P_2 = P \qquad (3.7)$$

The significance of Eq. (3.7) is so obvious that it may be overlooked. It is an equation which represents the vertical equilibrium not only of one particular portion of the transverse medium but of all portions, the longitudinal coordinate having disappeared. This reduces the problem of first harmonic responses of the bridge to a problem in beam deflection theory.

Equation (3.7) is conveniently regarded as the equilibrium equation per unit length of the transverse medium at midspan.

On considering deflections for the left-hand half of Fig. 3.3a, we have

$$-\left[\frac{(EI)_T}{L} \right] \delta x \, \frac{\partial^2 w(x, y)}{\partial y^2} = \left(P_1 \sin \frac{\pi x}{L} \right) \delta x \, y \qquad (3.8)$$

whence
$$-\left[\frac{(EI)_T}{L} \right] \frac{\partial^2 w(x, y)}{\partial y^2} = \left(P_1 \sin \frac{\pi x}{L} \right) y \qquad (3.9)$$

The deflection $w(x, y)$ can readily be written as

$$w(x, y) = w(y) \sin \frac{\pi x}{L} \qquad (3.10)$$

so that Eq. (3.9) becomes

$$-\left[\frac{(EI)_T}{L} \right] \frac{d^2 w(y)}{dy^2} = P_1 y \qquad (3.11)$$

Equation (3.11), like Eq. (3.7), is conveniently regarded as the deflection equation for a unit length of the transverse medium taken at midspan.

From now on the factors $\sin (\pi x/L)$ and δx will be omitted, and the deflection $w(y)$ will be written simply as w. Also, following the orthotropic plate notation, the transverse flexural stiffness per unit length $(EI)_T/L$ will be written as D_y. With these changes, Fig. 3.3a is replaced by Fig. 3.3b and Eq. (3.11) becomes

$$-D_y \frac{d^2 w}{dy^2} = P_1 y \qquad (3.12)$$

This equation is valid for $0 \leqslant y \leqslant S$ with the notation of Fig. 3.3.

By integrating Eq. (3.12) and inserting the condition that $dw/dy = 0$ when $y = S$, one obtains

$$-D_y \frac{dw}{dy} = \frac{P_1}{2} y^2 - \frac{P_1}{2} S^2 \tag{3.13}$$

By integrating a second time and inserting the condition that $w = a_1$ when $y = 0$, one obtains

$$-D_y w = \frac{P_1}{6} y^3 - \frac{P_1}{2} S^2 y - D_y a_1 \tag{3.14}$$

Finally, by putting $w = a_2$ when $y = S$,

$$-D_y a_2 = -\frac{P_1}{3} S^3 - D_y a_1 \tag{3.15}$$

Substituting for a_1 and a_2 from Eqs. (3.2) and (3.6) and multiplying throughout by $3/S^3$ then gives

$$P_1 \left(1 + \frac{3D_y}{kS^3} \right) = P_2 \left(\frac{3D_y}{kS^3} \right) \tag{3.16}$$

By recalling that D_y has been defined as $(EI)_T/L$ and k as $\pi^4 EI/L^4$, it is convenient to define a nondimensional parameter as

$$\eta = \frac{12D_y}{kS^3} \tag{3.17}$$

That is,

$$\eta = \frac{12}{\pi^4} \left(\frac{L}{S} \right)^3 \frac{(EI)_T}{EI} \tag{3.18}$$

Equations (3.7) and (3.18) are now solved, giving

$$\rho_1 = \frac{P_1}{P} = \frac{\eta}{4 + 3\eta}$$

$$\rho_2 = \frac{P_2}{P} = \frac{4 + \eta}{4 + 3\eta} \tag{3.19}$$

where ρ_1 and ρ_2 are distribution coefficients which give the fractions of the externally applied load accepted by girders 1 and 2 respectively.

It should be noted that these distribution coefficients apply with equal validity to load acceptance, bending moment, and deflection. For example, the total external load $P \sin (\pi x/L)$, if taken entirely by one girder, would give a free moment M_L, given by

$$M_L = \frac{PL^2}{\pi^2} \sin \frac{\pi x}{L} \qquad (3.20)$$

and a free deflection

$$w = \frac{PL^4}{\pi^4 EI} \sin \frac{\pi x}{L}$$

$$= A \sin \frac{\pi x}{L} \qquad (3.21)$$

For girder 1, we have

$$M_1 = \frac{P_1 L^2}{\pi^2} \sin \frac{\pi x}{L} \qquad (3.22)$$

and

$$w_1 = \frac{P_1 L^4}{\pi^4 EI} \sin \frac{\pi x}{L}$$

$$= a_1 \sin \frac{\pi x}{L} \qquad (3.23)$$

Comparing Eq. (3.20) with Eq. (3.22) and Eq. (3.21) with Eq. (3.23) gives

$$\frac{M_1}{M_L} = \frac{a_1}{A} = \frac{P_1}{P} = \rho_1 \qquad (3.24)$$

In practice, one works immediately in terms of the distribution of bending moments, shearing forces, etc. As is shown later, there is no difficulty in obtaining deflections also if these are needed.

The various distribution coefficients for a three-girder bridge with load on girder 1 are plotted in Fig. 3.10 against η.

Line load anywhere on the bridge. Figure 3.4a shows a longitudinal line load $P \sin (\pi x/L)$ at a distance eS from the left-hand girder. The deflections of the three girders are

$$a_1 \sin \frac{\pi x}{L} \qquad a_2 \sin \frac{\pi x}{L} \qquad a_3 \sin \frac{\pi x}{L}$$

Figure 3.4b shows the forces and deflections on the unit length of the transverse medium at midspan. Then,

$$-D_y \frac{d^2w}{dy^2} = P_1 y + P_2 [y - S] - P[y - eS] \qquad (3.25)$$

In Eq. (3.25) the bracketed terms are put equal to zero if their content becomes negative. By integrating Eq. (3.25) twice one obtains:

$$-D_y w = \frac{P_1}{6} y^3 + \frac{P_2}{6} [y - S]^3 - \frac{P}{6} [y - eS]^3 + C_1 y + C_2 \qquad (3.26)$$

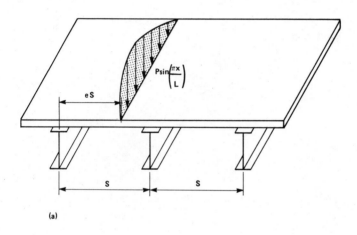

(a)

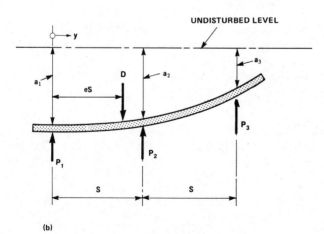

(b)

Figure 3.4 A three-girder bridge. (a) One line of sinusoidal loading. (b) Transverse slice of unit width.

Putting the deflection equal to a_1 at $y = 0$ gives the constant of integration C_2 as $-D_y a_1$. It is convenient to express this constant immediately as $-D_y(P_1/k)$ since equations in the unknown load acceptances P_1, P_2, and P_3 are being sought.

Now putting deflection as $a_2(=P_2/k)$ at $y = S$ and as $a_3(=P_3/k)$ at $y = 2S$ gives

$$-D_y \frac{P_2}{k} = \frac{P_1}{6} S^3 - \frac{P}{6} S^3[1 - e]^3 + C_1 S - D_y \frac{P_1}{k}$$

$$-D_y \frac{P_3}{k} = \frac{P_1}{6} 8S^3 + \frac{P_2}{6} S^3 - \frac{P}{6} S^3[2 - e]^3 + C_1 2S - D_y \frac{P_1}{k}$$

(3.27)

Elimination of the constant of integration C_1 between Eq. (3.27) in turn gives

$$-D_y(P_3 - 2P_2) = P_1 S^3 + \frac{P_2 S^3}{6}$$

$$- \frac{PS^3}{6} \times \{[2 - e]^3 - 2[1 - e]^3\} + D_y \frac{P_1}{k} \quad (3.28)$$

Whence, on multiplying through by $6/S^3$ and grouping terms,

$$P_1 \left(6 + \frac{\eta}{2}\right) + P_2(1 - \eta) + P_3 \left(\frac{\eta}{2}\right)$$

$$= P\{[2 - e]^3 - 2[1 - e]^3\} \quad (3.29)$$

Vertical equilibrium of the element of the transverse medium gives

$$P_1 + P_2 + P_3 = P \quad (3.30)$$

while, taking moments about the left-hand end, moment equilibrium gives

$$P_2 + 2P_3 = eP \quad (3.31)$$

Equations (3.29), (3.30), and (3.31) are sufficient to solve for P_1, P_2, and P_3. Conveniently, these three equations may each be divided through by P so as to work in terms of distribution coefficients ρ_1, ρ_2, and ρ_3. In this case, written in matrix form, one obtains

$$\begin{bmatrix} 1 & 1 & 1 \\ 0 & 1 & 2 \\ \left(6 + \dfrac{\eta}{2}\right) & (1 - \eta) & \dfrac{\eta}{2} \end{bmatrix} \begin{Bmatrix} \rho_1 \\ \rho_2 \\ \rho_3 \end{Bmatrix} = \begin{Bmatrix} 1 \\ e \\ [2 - e]^3 - 2[1 - e]^3 \end{Bmatrix} \quad (3.32)$$

This bank of equations is readily solved for any value of e. The distribution coefficients take very simple forms when the external loading is on a girder. For example, when the load is on girder 1, i.e., with $e = 0$, one finds

$$\rho_{11} = \frac{8 + 5\eta}{8 + 6\eta}$$

$$\rho_{21} = \frac{2\eta}{8 + 6\eta} \tag{3.33}$$

$$\rho_{31} = -\frac{\eta}{8 + 6\eta}$$

Again, with the load on girder 2, i.e., with $e = 1$, one finds

$$\rho_{12} = \frac{\eta}{4 + 3\eta}$$

$$\rho_{22} = \frac{4 + \eta}{4 + 3\eta} \tag{3.34}$$

$$\rho_{32} = \frac{\eta}{4 + 3\eta}$$

In Eqs. (3.33) and (3.34) a notation has been introduced for distribution coefficients which will be used consistently from now on. This is that ρ_{rs} means the fraction of load accepted by girder r when the external sinusoidal line loading is on girder S.

3.2 Second and Higher Harmonic Behavior in a Three-Girder Bridge

Figure 3.5 shows a line loading $P \sin (2\pi x/L)$ applied to the middle girder of a three-girder bridge. It is readily verified that all the conditions of equilibrium and compatibility may be satisfied by taking the line loadings accepted by the girders as $P_1 \sin (2\pi x/L)$, $P_2 \sin (2\pi x/L)$, and $P_3 \sin (2\pi x/L)$, where, because of symmetry, we have $P_3 = P_1$ in this case. Then all the reasoning of Sec. 3.1 remains valid, with the single change that $\pi X/L$ is being replaced by $2\pi x/L$. This means, in particular, that for second harmonic effects Eq. (3.18) is replaced by

$$\eta = \frac{12}{(2\pi)^4} \left(\frac{L}{S}\right)^3 \frac{(EI)_T}{EI} \tag{3.35}$$

and in general, for the pth harmonic, of form $\sin (p\pi x/L)$, we have

$$\eta = \frac{12}{(p\pi)^4} \left(\frac{L}{S}\right)^3 \frac{(EI)_T}{EI} \tag{3.36}$$

Thus, the η value for the second harmonic is one-sixteenth of that for the first harmonic, that for the third harmonic is one-eighty-first of that for the first harmonic, and so on. Provided only that the value of η is changed in this way from one harmonic to another, all the steps of analysis given by Eqs. (3.12) to (3.22) are still valid, as are Eqs. (3.33) and (3.34). This generalization, as will be seen later, is a key element in keeping bridge analysis within the scope of the personal computer.

3.3 Generalized Line Loading

The use of the method is illustrated with the help of a numerical example. Figure 3.6a shows the cross section of a slab-on-girder bridge with three girders. Figure 3.6b shows a plan view of the bridge which carries one line of wheels of a five-axle truck on the center girder. This loading is shown in Fig. 3.7a. The free bending moment M_L, using the harmonic analysis of beams given in Chap. 2, is given by

$$M_L = 1226.8 \sin \frac{\pi x}{L} + 61.7 \sin \frac{2\pi x}{L}$$

$$+ 36.5 \sin \frac{3\pi x}{L} - 14.4 \sin \frac{4\pi x}{L} + \cdots \tag{3.37}$$

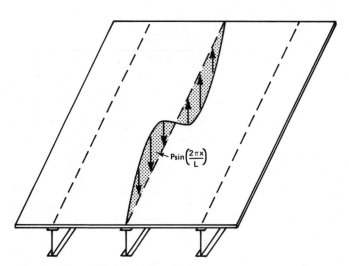

Figure 3.5 Second harmonic line load.

Figure 3.7b shows in the dashed line the bending-moment diagram for the load and, in the full line, the diagram given by the first two harmonics, i.e., a truncated form of M_L given by

$$M_L \simeq 1226.8 \sin \frac{\pi x}{L} + 61.7 \sin \frac{2\pi x}{L} \qquad (3.38)$$

The closeness of the actual bending-moment diagram and that obtained by Eq. (3.38), as shown in Fig. 3.7b, is noted. Ignoring torsional effects, one seeks to distribute this free bending moment between the three girders. This requires the calculation of the parameter η in accordance with Eq. (3.36). The first step is to calculate the total longitudinal bending stiffness of the bridge. It is well established [3] that the distribution characteristics of a bridge are not significantly changed

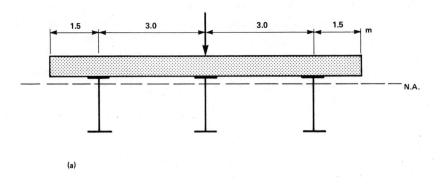

(a)

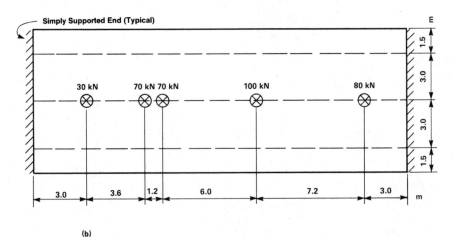

(b)

Figure 3.6 A three-girder bridge with one line of loads. (a) Cross section. (b) Plan.

by the phenomenon of shear lag. Hence, by referring to Fig. 3.6a, one takes the entire cross section, with a common neutral axis, and calculates its bending stiffness. By assuming a modular ratio of steel to concrete of 10.0, this is found to be $E_c \times 894 \times 10^9$ in terms of deck-slab concrete and mm⁴ units, where E_c is the modulus of elasticity of concrete.

Hence, we have the EI of each girder as

$$EI = \tfrac{1}{3}E_c \times 894 \times 10^9 = E_c \times 298 \times 10^9 \qquad (3.39)$$

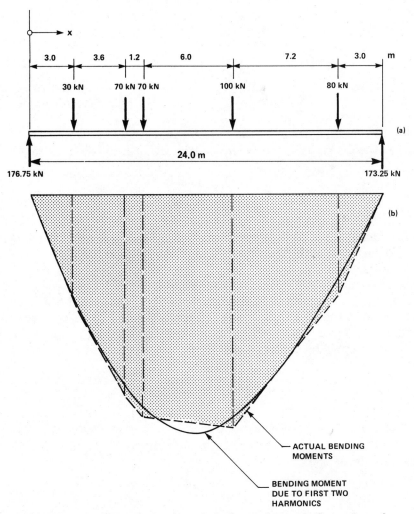

Figure 3.7 Free moments. (a) One line of wheels on a beam. (b) Comparison of true-moment diagram with that given by the first two harmonics.

In the transverse direction, we readily find

$$(EI)_T = E_c \times 16 \times 10^9 \tag{3.40}$$

Then for first harmonic effects, using Eq. (3.18) or Eq. (3.36), we find

$$\eta = \frac{12}{\pi^4} \left(\frac{24}{3}\right)^3 \frac{16}{298} = 3.39 \tag{3.41}$$

The distribution coefficients for the first harmonic are then obtained from Eq. (3.34) as

$$\rho_{12} = \rho_{32} = 0.239$$
$$\rho_{22} = 0.522 \tag{3.42}$$

The distribution coefficients given in Eq. (3.42) are applied to the first harmonic of the free-bending-moment diagram, i.e., to the term $1226.8 \sin (\pi x/L)$.

For the second harmonic the applicable value of η is $3.39/16 = 0.212$, so that the distribution coefficients, again using Eq. (3.34), become

$$\rho_{12} = \rho_{32} = 0.046$$
$$\rho_{22} = 0.908 \tag{3.43}$$

The distribution coefficients given in Eq. (3.43) are applied to the second harmonic of the free-bending-moment diagram, i.e., to the term $61.7 \sin (2\pi x/L)$.

By the time that the third harmonic is reached, the distribution coefficient ρ_{22} has risen to 0.980; i.e., virtually all the third harmonic is retained in the externally loaded girder, and similarly for all higher harmonics. Hence, with very good accuracy the bending moments M_1 and M_3, accepted by girders 1 and 3 respectively, are given by

$$M_1 = M_3 = 293.2 \sin \frac{\pi x}{L} + 2.8 \sin \frac{2\pi x}{L} \tag{3.44}$$

To obtain the bending moment M_2 accepted by girder 2, we note that the sum of the bending moments in the three girders is statically determinate, so that

$$M_1 + M_2 + M_3 = M_L \tag{3.45}$$

Hence,

$$M_2 = M_L - (M_1 + M_3) \tag{3.46}$$

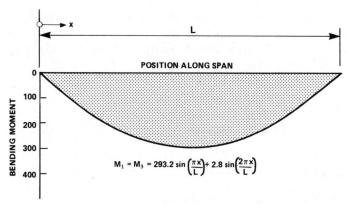

Figure 3.8   Bending-moment diagram for girders 1 and 3.

which, using Eq. (3.44), gives

$$M_2 = M_L - \left(586.4 \sin \frac{\pi x}{L} + 5.6 \sin \frac{2\pi x}{L}\right) \qquad (3.47)$$

It is much more convenient to express M_2 in the manner of Eq. (3.47) than by a harmonic series which involves many terms. Equations (3.44) and (3.47) make it clear that the required bending moments M_1, M_2, and M_3 can be conveniently expressed in terms of (1) the polygonal free-bending-moment diagram M_L and (2) distributions of only the first two harmonics.

Figure 3.8 shows the bending-moment diagram for girders 1 and 3, while Fig. 3.9 shows that for girder 2. It will be noted that in this

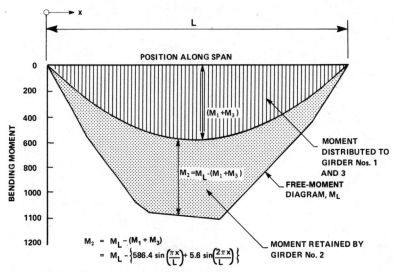

Figure 3.9   Bending moment retained by girder 2.

example even the second harmonic distribution is not significant, so that to design accuracy one could distribute only the first harmonic of M_L and retain all the other harmonics fully in the loaded girder. The eventual bending-moment diagrams of Figs. 3.8 and 3.9 would be imperceptibly different from those shown.

Figure 3.10 shows in graphical form the distribution coefficients ρ_{11}, ρ_{21}, and ρ_{31} plotted against η. It is noted that with the generalized definition of η given in Eq. (3.36) these curves are available for all harmonics.

3.4   General Bridge with *N* Girders

First harmonic effects.   By adopting an approach similar to that of Sec. 3.1, an analytical method is now developed for the general case. The

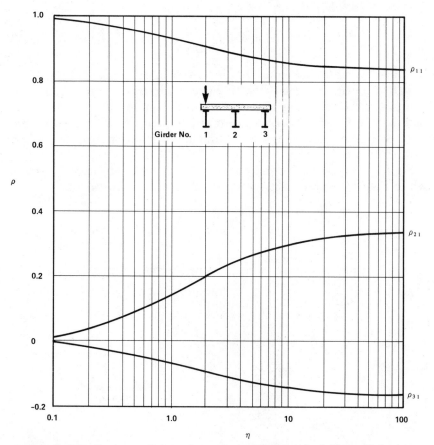

Figure 3.10   Distribution coefficients for torsionless three-girder bridge with load on girder 2.

bridge consists of N girders having possibly different flexural stiffnesses $(EI)_1, (EI)_2, \ldots, (EI)_N$. The girder spacings are $S_1, S_2, \ldots, S_{N-1}$, and the distances of girders 2 to N from the left-hand side of the cross section are $b_1, b_2, \ldots, b_{N-1}$ respectively. The bridge carries a line loading $P \sin(\pi x/L)$ at a distance b_e from the left-hand girder, as shown in Fig. 3.11. We define

$$k_m = \frac{\pi^4 (EI)_m}{L^4} \tag{3.48}$$

and

$$\eta_m = \frac{12}{\pi^4} \left(\frac{L}{S_m}\right)^3 \frac{(EI)_T}{(EI)_m} \tag{3.49}$$

It may be noted that the definition of η_N requires a value for S_N, which does not exist physically. We arbitrarily define $S_N = S_{N-1}$ for convenience in calculation. The quantity S_N eventually disappears from the calculations; it is given a value simply so that generally applicable formulas may be given to the computer.

The moment-curvature relationship for the unit length of transverse

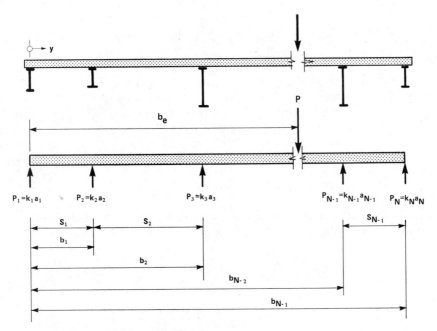

Figure 3.11 General bridge with N girders.

medium at midspan is

$$-D_y \frac{d^2w}{dy^2} = P_1 y + P_2[y - b_1] + P_3[y - b_2]$$

$$+ \cdots + P_{N-1}[y - b_{N-2}] - P[y - b_e] \quad (3.50)$$

where the bracketed terms are put equal to zero if their content is negative.

Integrating this equation twice gives

$$-D_y w = \frac{P_1}{6} y^3 + \frac{P_2}{6}[y - b_1]^3 + \frac{P_3}{6}[y - b_2]^3$$

$$+ \cdots + \frac{P_{N-1}}{6}[y - b_{N-2}]^3 - \frac{P}{6}[y - b_e]^3 + C_1 y + C_2 \quad (3.51)$$

The constant of integration C_2 is immediately found to be $-D_y a_1[= -D_y(P_1/k_1)]$ from the condition that $w = a_1$ at $y = 0$. By putting $w = a_2$ at $y = b_1$ there results

$$-D_y a_2 \left(= -D_y \frac{P_2}{k_2} \right) = \frac{P_1}{6} b_1{}^3 - \frac{P}{6}[b_1 - b_e]^3 + C_1 b_1 - D_y \frac{P_1}{k_1} \quad (3.52)$$

which gives

$$C_1 = \frac{b_1{}^2}{6} \left\{ -P_1 \left(1 + \frac{\eta_1}{2} \right) - P_2 \left(\frac{k_1}{k_2} \right) \frac{\eta_1}{2} + P \left[\frac{b_1 - b_e}{S_1} \right]^3 \right\} \quad (3.53)$$

Now putting in succession $w = a_3, a_4, \ldots, a_N$ at $y = b_2, b_3, \ldots, b_{N-1}$ respectively gives $(N - 2)$ equations for the unknowns $P_1, P_2, \ldots, P_N$. The remaining two equations are obtained from vertical equilibrium and moment equilibrium. They are

$$P_1 + P_2 + P_3 + \cdots + P_N = P \quad (3.54)$$

$$P_2 b_1 + P_3 b_2 + \cdots + P_N b_{N-1} = P b_e \quad (3.55)$$

The $(N - 2)$ equations referred to have a pattern. Considering the situation at $y = b_{r-1}$, that is, at the position of girder r, we have

$$-D_y \frac{P_r}{k_r} = \frac{P_1}{6} b_{r-1}^3 + \frac{P_2}{6}[b_{r-1} - b_1]^3 + \cdots + \frac{P_{r-1}}{6}[b_{r-1} - b_{r-2}]^3$$

$$- \frac{P}{6}[b_{r-1} - b_e]^3 + C_1 b_{r-1} + C_2 \quad (3.56)$$

Noting that $(b_{r-1} - b_{r-2})$ is equal to S_{r-1}, we multiply Eq. (3.55) throughout by $6/S_{r-1}^3$, substitute for C_1 and C_2, and gather up terms. Then, for $r = 3$ to N,

$$P_1 \left\{ -\left(1 + \frac{\eta_1}{2}\right) \left(\frac{b_1}{S_{r-1}}\right)^2 \left(\frac{b_{r-1}}{S_{r-1}}\right) + \left(\frac{b_{r-1}}{S_{r-1}}\right)^2 - \frac{\eta_{r-1}}{2} \left(\frac{k_{r-1}}{k_r}\right) \right\}$$

$$+ P_2 \left\{ -\frac{\eta_1}{2} \left(\frac{b_1}{S_{r-1}}\right)^2 \left(\frac{b_{r-1}}{S_{r-1}}\right) \left(\frac{k_1}{k_2}\right) + \left[\frac{b_{r-1} - b_1}{S_{r-1}}\right]^3 \right\}$$

$$+ \sum_{g=2}^{g=r-2} P_{g+1} \left[\frac{b_{r-1} - b_g}{S_{r-1}}\right]^3 + P_r \left(\frac{\eta_{r-1}}{2}\right) \left(\frac{k_{r-1}}{k_r}\right) = P \left[\frac{b_{r-1} - b_e}{S_{r-1}}\right]^3$$

$$(3.57)$$

The set of N equations given by Eqs. (3.54), (3.55), and (3.57) is readily solved by computer. As noted earlier, it is usually convenient to divide each of the N equations by P so that they become equations in the distribution coefficients ρ_1, ρ_2 etc.

Second and higher harmonic effects. For the distribution of nth harmonic effects it is only necessary to write $n\pi$ in place of π in the formulas for k_m and η_m, i.e., Eqs. (3.48) and (3.49). Once this change has been made, the N simultaneous Eqs. (3.54), (3.55), and (3.57) given above remain valid.

3.5 Transverse Moments

Once the values of $P_1, P_2, \ldots, P_N$ have been established, the transverse moments follow directly. It is only necessary to substitute these values into Eq. (3.50), and an expression results for the transverse moments per unit length of the transverse medium at any position y across the cross section. Transverse moments due to the first, second, etc., harmonics are readily superposed.

3.6 Variation of Distribution Effects along the Length of the Bridge

Inspection of bending-moment diagrams such as those shown in Fig. 3.9 demonstrate clearly that the fractions M_1/M_L, M_2/M_L, etc., of the free bending moment M_L taken by individual girders are different at different longitudinal positions of the bridge. This difference arises because the successive harmonic components of M_L are distributed between the girders in different proportions. Thus the bending-moment diagrams M_1, M_2, etc., are in general different in shape from one another, and all of them are different in shape from the free bending moment M_L.

Most distribution coefficient methods (e.g., those developed in Refs. 1 and 4) are based upon the simplifying assumption that fractions such as M_1/M_L and M_2/M_L are constant all along the length of the bridge. In effect, this assumes that M_L is purely first harmonic, which is seldom or never the case. It is not difficult to show that even if bending moments in a bridge are fairly close to being purely first harmonic the same cannot be said of shear forces. It is for this reason that the distribution coefficient methods mentioned above cannot be relied upon to give realistic distributions of shear forces.

It is an advantage of the semicontinuum approach that the need to assume constancy of fractions such as M_1/M_L and M_2/M_L is obviated. This means that the method, although very simple to use, is in fact a refined method of analysis and provides accuracy comparable with that of other refined methods.

REFERENCES

1. Cusens, A. R., and Pama, R. P.: *Bridge Deck Analysis,* Wiley, London, 1975, pp. 85–132.
2. Hendry, A. W., and Jaeger, L. G.: *The Analysis of Grid Frameworks and Related Structures,* Prentice-Hall, Englewood Cliffs, N.J., 1958.
3. Jaeger, L. G., and Bakht, B.: The grillage analogy in bridge analysis, *Canadian Journal of Civil Engineering,* 9(2), 1982, pp. 224–235.
4. Rowe, R. E.: *Concrete Bridge Design,* Applied Science Publishers, London, 1972, pp. 62–114.

General Semicontinuum Method

The previous semicontinuum method [3] was developed mainly for bridges with negligible torsional rigidities; this method included an approximate way of taking account of longitudinal torsional stiffnesses. A generalized modification of this method, which can now take account of nonzero values of torsional rigidities in both the longitudinal and the transverse directions, is presented in this chapter for slab-on-girder bridges. The method now presented has the further generalizing feature that it can deal with girders of unequal flexural and torsional rigidities at unequal spacings. Only slab-on-girder bridges with right spans and simple supports are dealt with in this chapter. The methods of analysis for continuous bridges are given in Chap. 5, methods for skew bridges in Chap. 6, and those for rigid frame bridges in Chap. 8. The application of the general method to other than slab-on-girder bridges is dealt with in Chap. 7.

4.1 Significance of Harmonic Load Representation

The significance of representation of loads in the semicontinuum method has been briefly dealt with in Chap. 3. This topic is discussed in detail here because it is basic to an understanding of the method.

The example of four girders, or longitudinal beams, is taken as a vehicle for discussion. The four girders are first considered to be connected by three transverse beams, and the second girder is subjected to a point load at the midspan, as shown in Fig. 4.1. As discussed in Sec. 1.1.3, the various girders will receive different patterns of loads. For example, as shown in Fig. 4.1, girder 1 will accept loading in the form of three downward-acting point loads. Because of different patterns of loadings accepted by the various girders, the deflection patterns of the girders will differ from each other. The analysis of the problem shown in Fig. 4.1 will require the solution of 36 unknowns. If the number of transverse members is increased to seven, as would be re-

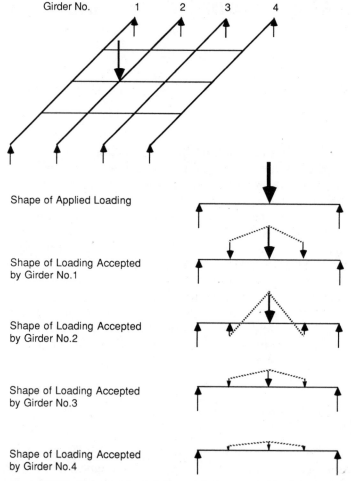

Figure 4.1 Distribution of loads in four girders connected by three transverse beams.

quired for a realistic grillage idealization, the number of unknowns will be increased to 84.

The number of transverse beams is now greatly increased and, as shown in Fig. 4.2, the second girder is subjected to a distributed load which is in the shape of a half sine wave. This loading, as discussed in Chap. 2, corresponds to the first term of a harmonic series representing a point load. For the case under consideration, all the loadings accepted by the various girders will have the same shape. Consequently, the deflection profiles of the girders will have the same shapes and will be related to each other by scalar multipliers. The implication of similar deflection patterns of the girders is that the assembly of beams can now be *exactly* analyzed by considering only a transverse slice of the structure. In this way, the number of unknowns for the four-girder bridge is reduced to eight without compromising accuracy. In fact, accuracy is enhanced by considering an infinite number of transverse members as compared with the seven (or some such small number) that are employed in the usual grillage idealization.

The distribution of loading represented by the second term of the harmonic series is shown in Fig. 4.3. As can be seen in the figure, the loads accepted by the various girders in this case are also of similar shapes, with the result that the problem of distribution of this loading

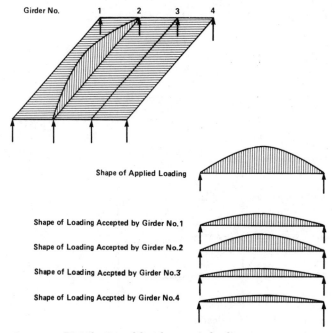

Figure 4.2 Distribution of first harmonic loading.

and those represented by higher harmonics can again be solved by determining only eight unknowns.

It is noted that the various harmonics are distributed among the girders in different ways, with the result that when the distributed loads are added up, the total loads accepted by the girders are not, in general, similar to one another. This phenomenon is illustrated in Fig. 4.4 for a bridge with four girders and carrying a point load at the midspan of girder 2. For a point load at midspan, the even-numbered harmonic terms of the harmonic series are all equal to zero, and the load is represented by only the odd-numbered terms. It can be seen in Fig. 4.4 that the sums of the loads accepted by the various girders do not have similar shapes.

The key to the efficiency of the semicontinuum method lies in the separation of the dissimilar loads accepted by the various girders into components of similar shapes, as a result of which the number of unknowns is dramatically reduced.

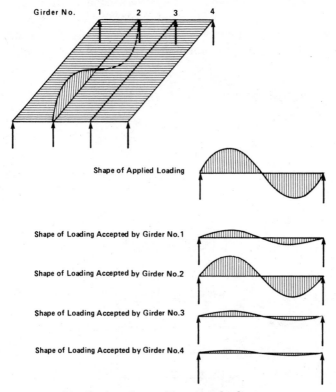

Figure 4.3 Distribution of second harmonic loading.

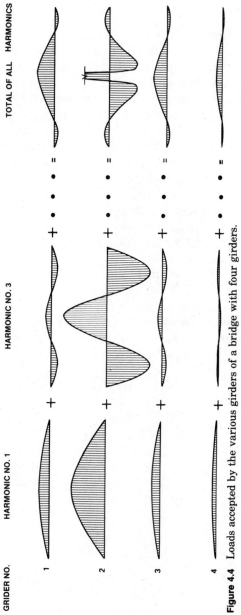

Figure 4.4 Loads accepted by the various girders of a bridge with four girders.

4.2 First Harmonic Relationships

As was the case in Chap. 3, the behavior of the bridge is first examined under pure first harmonic conditions; i.e., an externally applied line load in the shape of a half sine wave is applied, as shown in Fig. 4.5. It is then readily verified, as shown below, that even in the presence of torsional effects the vertical loads accepted by all longitudinal girders have this same half sine waveshape, as does the distribution of twisting moments along the girders (which are also the distributed bending moments in the transverse medium). Yet again, there is a distribution of twisting moments in the transverse medium which, when applied to the longitudinal girders, is equivalent to distributed loadings with the same half sine waveform.

Figure 4.6 shows longitudinal girders 1 and 2 and the portion of transverse medium between them. The girders are of length L and are at a center-to-center distance S_1 apart. A coordinate x is measured in the longitudinal direction of the bridge.

The externally applied line load shown in Fig. 4.5 is of intensity $P \sin (\pi x/L)$.

We examine what interactions of force and moment are needed between the girders and the transverse medium in order that girder 1 may deflect into a pure sine wave of form

$$w_1 = a_1 \sin \frac{\pi x}{L} \tag{4.1}$$

and similarly for girder 2,

$$w_2 = a_2 \sin \frac{\pi x}{L} \tag{4.2}$$

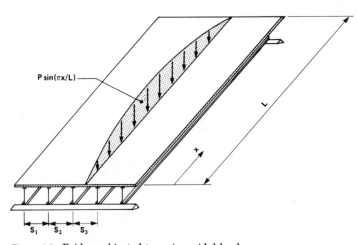

$P \sin(\pi x/L)$

S_1 S_2 S_3

Figure 4.5 Bridge subjected to a sinusoidal load.

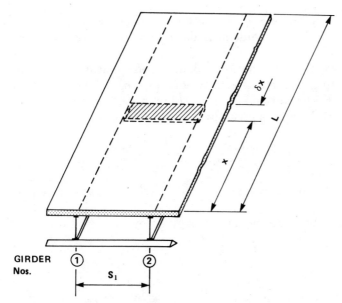

Figure 4.6 Segment of a bridge.

where, as shown in Fig. 4.7, w is vertical deflection, a is its value at midspan (i.e., its amplitude), and the subscripts refer to girder numbers. At the same time, girder 1 is permitted to rotate about its own longitudinal centerline through an angle

$$\phi_1 = v_1 \sin \frac{\pi x}{L} \tag{4.3}$$

and similarly for girder 2,

$$\phi_2 = v_2 \sin \frac{\pi x}{L} \tag{4.4}$$

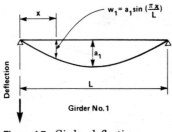
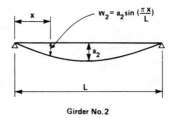

Figure 4.7 Girder deflections.

where, as shown in Fig. 4.8, ϕ is rotation from the initially vertical position of a girder, v is its value at midspan, and the subscripts refer to girder numbers.

Figure 4.9 shows the combination of distributed vertical loading and distributed twisting moment that is needed to achieve the deflection $a_m \sin (\pi x/L)$ and the rotation $v_m \sin (\pi x/L)$ in girder m. The intensity of the vertical line load p_x is given by

$$p_x = \frac{(EI)_m \pi^4}{L^4} a_m \sin \frac{\pi x}{L} \qquad (4.5)$$

This line load is supported by concentrated reactions at each end equal to $[(EI)_m \pi^3] a_m / L^3$.

The intensity of twisting moment t_x is given by

$$t_x = \frac{(GJ)_m \pi^2}{L^2} v_m \sin \frac{\pi x}{L} \qquad (4.6)$$

This distribution of twisting moment is resisted by concentrated twisting moments $[(GJ)_m \pi] v_m / L$ at each end.

It is noted that in this pure first harmonic behavior the local intensity of vertical load is proportional to the local value of deflection and that the local intensity of twisting moment per unit length is proportional to the local value of rotation.

The next step is to consider how these distributions of vertical load and twisting moment come to be applied to the girders from the transverse medium.

Figure 4.10 shows the strip of transverse medium of width δx contained between values x and $(x + \delta x)$ of the longitudinal coordinate. The slopes of girders 1 and 2 at the ends of this strip are $(\pi/L)a_1 \cos (\pi x/L)$ and $(\pi/L)a_2 \cos (\pi x/L)$ respectively. Since the torsional rigidity of this strip is $D_{yx} \delta x$, it follows that the equal and opposite twisting moments at the ends of the strip are of magnitude $(D_{yx}/S_1)(\pi/L)(a_2 - a_1) \cos (\pi x/L) \delta x$. These twisting moments can each be rep-

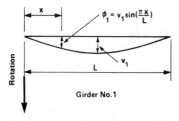

Figure 4.8 Girder rotations.

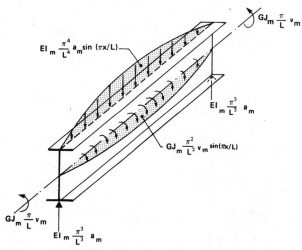

Figure 4.9 Forces on girder m due to first harmonic loading.

resented as a pair of forces of magnitude $(D_{yx}/S_1)(\pi/L)(a_2 - a_1)$ $\cos(\pi x/L)$ parallel to one another, but in opposite senses, and a distance δx apart, as shown in Fig. 4.11. It is now a simple matter to use the technique introduced many years ago by Thompson and Tait [9] and replace all the twisting moments on all the strips of transverse medium by such pairs of forces, the net result of which is a downward loading

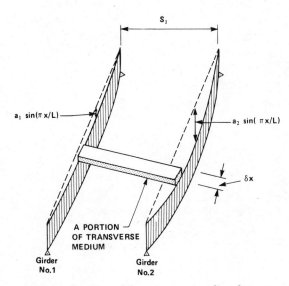

Figure 4.10 Segment of the transverse medium between two adjacent girders.

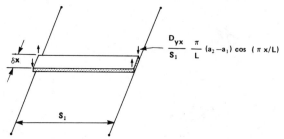

Figure 4.11 Forces on a segment of the transverse medium.

on girder 1 of intensity R_1, which is given by

$$R_1 = \frac{D_{yx}}{S_1} \frac{\pi^2}{L^2} (a_2 - a_1) \sin \frac{\pi x}{L} \tag{4.7}$$

and a downward loading on girder 2 of intensity R_2, given by

$$R_2 = \frac{D_{yx}}{S_1} \frac{\pi^2}{L^2} (a_1 - a_2) \sin \frac{\pi x}{L} \tag{4.8}$$

Appendix II provides details of the method by Thompson and Tait by which distributed twisting moments can be replaced by distributed vertical loading. Replacement of transverse twisting moments in the deck slab by vertical loading on girders is diagrammatically shown in Fig. 4.12.

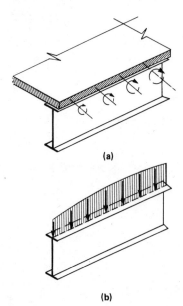

(a)

(b)

Figure 4.12 Dealing with twisting moments in the deck slab. (a) Girder segment subjected to twisting moment from the deck slab. (b) Deck-slab twisting moments replaced by vertical loading.

Let girder 1 be an outside girder and girder 2 an inside girder. Then, from Eqs. (4.5) and (4.7) the vertical interactive line force between girder 1 and the transverse medium is

$$P_1 \sin \frac{\pi x}{L} = \frac{(EI)_1 \pi^4}{L^4} a_1 \sin \frac{\pi x}{L} - \frac{D_{yx}}{S_1} \frac{\pi^2}{L^2} (a_2 - a_1) \sin \frac{\pi x}{L} \quad (4.9)$$

Hence the amplitude P_1 of the vertical interactive force between girder 1 and the transverse medium is given by

$$P_1 = \frac{(EI)_1 \pi^4}{L^4} a_1 - \frac{D_{yx}}{S_1} \frac{\pi^2}{L^2} (a_2 - a_1) \quad (4.10)$$

Insofar as girder 2 is concerned, there are twisting moments in the transverse medium on either side of it. The portion of the medium between girders 1 and 2 gives the line load of Eq. (4.8), and similarly the portion between girders 2 and 3 gives a line load of intensity $(D_{yx}\pi^2/S_2 L^2)(a_3 - a_2) \sin (\pi x/L)$. Hence

$$P_2 \sin \frac{\pi x}{L} = \frac{(EI)_2 \pi^4}{L^4} a_2 \sin \frac{\pi x}{L} - \frac{D_{yx}}{S_1} \frac{\pi^2}{L^2} (a_1 - a_2) \sin \frac{\pi x}{L}$$

$$- \frac{D_{yx}}{S_2} \frac{\pi^2}{L^2} (a_3 - a_2) \sin \frac{\pi x}{L} \quad (4.11)$$

whence $P_2 = \dfrac{(EI)_2 \pi^4}{L^4} a_2 + \dfrac{D_{yx}}{S_1} \dfrac{\pi^2}{L^2} (a_2 - a_1) - \dfrac{D_{yx}}{S_2} \dfrac{\pi^2}{L^2} (a_3 - a_2)$

$$(4.12)$$

Equation (4.12) is the typical equation for an interior girder. Clearly, if girder r is an interior one, we have

$$P_r = \frac{(EI)_r \pi^4}{L^4} a_r + \frac{D_{yx}}{S_{r-1}} \frac{\pi^2}{L^2} (a_r - a_{r-1}) - \frac{D_{yx}}{S_r} \frac{\pi^2}{L^2} (a_{r+1} - a_r) \quad (4.13)$$

For the right-hand (Nth) girder we have

$$P_N = \frac{(EI)_N \pi^4}{L^4} a_N + \frac{D_{yx}}{S_{N-1}} (a_N - a_{N-1}) \quad (4.14)$$

We now consider the forces and moments that are applied to a strip of transverse medium between x and $(x + \delta x)$ and the resulting bending deflections. The strip has bending stiffness $D_y \, \delta x$.

At the position of girder 1, the strip of medium experiences an upward force of size $P_1 \sin (\pi x/L) \, \delta x$, where P_1 is given by Eq. (4.10), and a bending moment (which is positive anticlockwise if rotations are accounted positive clockwise) of size $[(GJ)_1 \pi^2 v_1]/L^2 \sin (\pi x/L) \, \delta x$ in accordance with Eq. (4.6). The deflection and rotation at the position of girder 1 are $a_1 \sin (\pi x/L)$ and $v_1 \sin (\pi x/L)$ respectively. The other girders give similar results.

For brevity, let

$$m_1 = \frac{(GJ)_1 \pi^2}{L^2}$$

$$m_2 = \frac{(GJ)_2 \pi^2}{L^2} \qquad (4.15)$$

$$\vdots$$

Then on canceling a common factor of $\sin (\pi x/L)$ between all forces, moments, deflections, and rotations and on canceling a common factor δx between the applied forces and moments and the bending stiffness of the strip, the reactive forces on a transverse slice of unit width at the midspan are found to be as shown in Fig. 4.13.

It is noted that Fig. 4.13 is representative of the behavior per unit

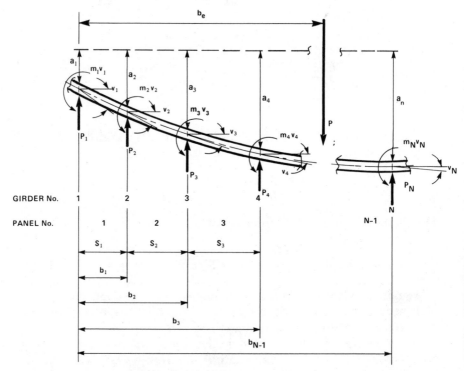

Figure 4.13 Responses at midspan in a bridge with N girders.

length of all strips of the transverse medium, on multiplying through-out by sin $(\pi x/L)$.

It will be apparent that the problem of distributing first harmonic effects in the bridge is thus reduced to solving the problem of the deflection of a beam supported by a system of linear springs. Specifically, with reference to Fig. 4.13, an external load of size P is shared between the girders as $P_1, P_2, \ldots, P_N$. Further, the beam has deflections $a_1, a_2, \ldots, a_N$ and rotations $v_1, v_2, \ldots, v_N$ in response to (1) the external force P; (2) support forces $P_1, P_2, \ldots, P_N$, which are given by Eqs. (4.10), (4.12), etc.; and (3) restraint moments $m_1v_1, m_2v_2, \ldots, m_Nv_N$, where m_1, $m_2, \ldots, m_N$ are given by Eq. (4.15).

As an example, the representation of a transverse strip of a five-girder bridge by a beam supported on a system of springs is shown in Fig. 4.14. The cross section of the bridge is shown in Fig. 4.14a. In Fig. 4.14b, the flexural and torsional rigidities of the girders are replaced by springs of stiffnesses k_r and m_r respectively, with subscripts referring to girder numbers. The analogy is now a unit length of deck slab resting on elastic springs. For girder r, the expressions for k_r and m_r are

$$k_r = \frac{\pi^4}{L^4}(EI)_r \tag{4.16}$$

$$m_r = \frac{\pi^2}{L^2}(GJ)_r \tag{4.17}$$

The transverse torsional rigidity of the transverse medium, i.e., the deck slab, is represented by springs of stiffness c_r with the subscript referring to the panel number, identified in Fig. 4.14a. As may be seen from Eqs. (4.10), (4.12), etc., the expression for this spring stiffness for panel r is

$$c_r = \frac{\pi^2}{L^2}\frac{D_{yx}}{S_r} \tag{4.18}$$

The analogy now consists of a system of springs shown in Fig. 4.14c. The same logic used for the first harmonic can now be used for higher harmonics.

4.3   Derivation of Equations

As a preliminary to setting up the definitive equations for the general case, the following definitions are made for harmonic 1. Some of these quantities have already been defined in Sec. 4.2, but they are repeated here so that the list now given may be complete.

$$g = \frac{\pi^2 D_{yx}}{L^2}$$

$$k_r = \frac{\pi^4 (EI)_r}{L^4} \qquad \text{for} \qquad r = 1, 2, \ldots, N$$

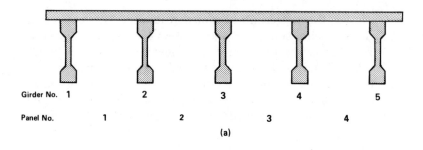

Girder No.   1   2   3   4   5

Panel No.   1   2   3   4

(a)

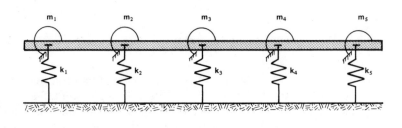

(b)

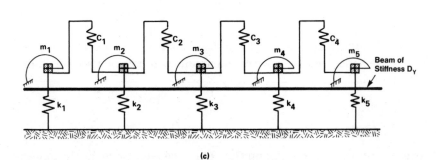

(c)

Figure 4.14 Transformation of a bridge into a system of springs. (a) Actual cross section. (b) Deck slab on springs. (c) A system of springs.

$$m_r = \frac{\pi^2 (GJ)_r}{L^2} \qquad \text{for} \qquad r = 1, 2, \dots, N$$

$$c_r = \frac{g}{S_r} = \frac{\pi^2}{L^2} \frac{D_{yx}}{S_r} \qquad \text{for} \qquad r = 1, 2, \dots, (N-1)$$

$$\eta_r = \frac{12 D_y}{k_r S_r^3} = \frac{12}{\pi^4} \left(\frac{L}{S_r}\right)^3 \frac{L D_y}{(EI)_r} \qquad \text{for} \qquad r = 1, 2, \dots, (N-1)$$

$$\lambda_r = \frac{c_r}{k_r} = \frac{1}{\pi^2} \left(\frac{L}{S_r}\right)^2 \frac{S_r D_{yx}}{(EI)_r} \qquad \text{for} \qquad r = 1, 2, \dots, (N-1)$$

$$\mu_r = \frac{m_r}{k S_r^2} = \frac{1}{\pi^2} \left(\frac{L}{S_r}\right)^2 \frac{(GJ)_r}{(EI)_r} \qquad \text{for} \qquad r = 1, 2, \dots, (N-1)$$

$$(4.19)$$

For harmonic m, π in the above equations is replaced by $m\pi$.

It is important to note that the valid range of the last four quantities is $r = 1$ to $(N-1)$ only. This is true because they involve the spacings S_r between girders, of which there are $(N-1)$. For computation purposes, it is very convenient to have definitions which extend up to $r = N$, and great care must be exercised in how this is done.

Insofar as the definitions of c_r, η_r, and μ_r are concerned, we define $S_N = S_{N-1}$ and then immediately extend the validity of the formulas given to be from $r = 1$ to $r = N$. However, the same cannot be done for λ_r; in this case, in order to have generally applicable formulas we must define $\lambda_N = 0$. Further, and still with the objective of having generally applicable formulas for the computer, we define

$$b_0 = 0$$
$$\lambda_0 = 0$$
$$k_0 = 0 \qquad\qquad (4.20)$$
$$S_0 = 0$$

By using these definitions, Eqs. (4.10), (4.13), and (4.14) may be rewritten:

$$P_1 = k_1 a_1 - c_1 (a_2 - a_1) = k_1 [(1 + \lambda_1) a_1 - \lambda_1 a_2] \qquad (4.21)$$

$$P_r = k_r a_r + c_{r-1}(a_r - a_{r-1}) - c_r(a_{r+1} - a_r)$$

$$= k_r \left[-\frac{c_{r-1}}{k_r} a_{r-1} + \left(1 + \frac{c_{r-1}}{k_r} + \lambda_r\right) a_r - \lambda_r a_{r+1} \right] \qquad (4.22)$$

$$P_N = k_N a_N + c_{N-1}(a_N - a_{N-1})$$

$$= k_N \left[-\frac{c_{N-1}}{k_N} a_{N-1} + \left(1 + \frac{c_{N-1}}{k_N} \right) a_N \right] \qquad (4.23)$$

where Eq. (4.22) is available for $r = 2, 3, \ldots, (N - 1)$.

Referring again to Fig. 4.13, vertical equilibrium of the whole gives

$$P_1 + P_2 + \cdots + P_N = P \qquad (4.24)$$

On substituting for $P_1, P_2, \ldots, P_N$ from Eqs. (4.21), (4.22), and (4.23), this vertical-equilibrium equation may be written as

$$k_1 a_1 + k_2 a_2 + \cdots + k_N a_N = P \qquad (4.25)$$

Taking moments about the left-hand girder gives

$$(P_2 b_1 + P_3 b_2 + \cdots + P_N b_{N-1})$$
$$+ (m_1 v_1 + m_2 v_2 + \cdots + m_N v_N) = P b_e \qquad (4.26)$$

On substituting for $P_2, P_3, \ldots, P_N$ from Eqs. (4.12), (4.13), and (4.14) and for $m_1, m_2, \ldots, m_N$ from Eqs. (4.15), this moment equilibrium equation may be written

$$\left(-\frac{g}{k_1} \right) k_1 a_1 + \sum_{r=2}^{r=N-1} b_{r-1} k_r a_r + \left(b_{N-1} + \frac{g}{k_N} \right) k_N a_N$$

$$+ \sum_{r=1}^{r=N} (\mu_r S_r) k_r S_r v_r = P b_e \qquad (4.27)$$

For purposes of calculation, it is convenient to divide Eq. (4.27) on both sides by b_{N-1}; it is worthy of note in this equation that the $2N$ unknowns are $k_1 a_1, k_2 a_2, \ldots, k_N a_N$ and $k_1 S_1 v_1, k_2 S_2 v_2, \ldots, k_N S_N v_N$. This is a convenient formulation.

The bending-moment–curvature relationship gives

$$-D_y \frac{d^2 w}{dy^2} = P_1 y + P_2[y - b_1] + \cdots + P_{N-1}[y - b_{N-2}]$$

$$- P[y - b_e] - m_1 v_1 - [m_2 v_2] - \cdots - [m_{N-1} v_{N-1}] \qquad (4.28)$$

In Eq. (4.28) the term $[m_2 v_2]$ is ignored if $y < b_1$, the term $[m_3 v_3]$ is ignored if $y < b_2$, and so on. Integrating this equation and inserting the condition that $dw/dy = v_1$ when $y = 0$ give

$$-D_y \frac{dw}{dy} = \frac{P_1}{2} y^2 + \frac{P_2}{2} [y - b_1]^2$$

$$+ \cdots + \frac{P_{N-1}}{2} [y - b_{N-2}]^2 - \frac{P}{2} [y - b_e]^2 - m_1 v_1 y - m_2 v_2 [y - b_1]$$

$$- \cdots - m_{N-1} v_{N-1} [y - b_{N-2}] - D_y v_1 \quad (4.29)$$

By substituting $y = b_p$ in Eq. (4.29), one obtains an equation valid at the position of the $(p + 1)$th girder. Then, by substituting for P_1, $P_2, \ldots, P_{N-1}$ and $m_1, m_2, \ldots, m_N$ and then multiplying throughout by $2/S_p^2$, there results

$$\left\{ \left(\frac{b_p}{S_p}\right)^2 (1 + \lambda_1) - \left(\frac{b_p - b_1}{S_p}\right)^2 \lambda_1 \right\} k_1 a_1$$

$$+ \sum_{t=2}^{t=p} \left\{ -\left(\frac{b_p - b_{t-2}}{S_p}\right)^2 \frac{c_{t-1}}{k_t} \right.$$

$$+ \left(\frac{b_p - b_{t-1}}{S_p}\right)^2 \left(1 + \frac{c_{t-1}}{k_t} + \lambda_t\right) - \left(\frac{b_p - b_t}{S_p}\right)^2 \lambda_t \right\} k_t a_t$$

$$- \left(\frac{\lambda_p k_p}{k_{p+1}}\right) k_{p+1} a_{p+1}$$

$$- \left\{ \frac{\eta_1}{6} \left(\frac{S_1}{S_p}\right)^2 + 2\mu_1 \left(\frac{S_1}{S_p}\right) \left(\frac{b_p}{S_p}\right) \right\} k_1 S_1 v_1$$

$$- \sum_{t=2}^{t=p} \left\{ 2\mu_t \left(\frac{S_t}{S_p}\right) \left(\frac{b_p - b_{t-1}}{S_p}\right) \right\} k_t S_t v_t$$

$$+ \left(\frac{\eta_{p+1}}{6}\right) \left(\frac{S_{p+1}}{S_p}\right)^2 k_{p+1} S_{p+1} v_{p+1}$$

$$= P \left[\frac{b_p - b_e}{S_p}\right]^2 \quad (4.30)$$

Equation (4.30) is available for $p = 1, 2, \ldots, (N - 1)$. It is noted that for $p = 1$ the terms which involve summation over t for values from 2 to p are, of course, absent.

By integrating Eq. (4.29) and inserting the condition that $w = a_1$

when $y = 0$, there results

$$-D_y w = \frac{P_1}{6} y^3 + \frac{P_2}{6} [y - b_1]^3$$

$$+ \cdots + \frac{P_{N-1}}{6} [y - b_{N-2}]^3 - \frac{P}{6} [y - b_e]^3 - \frac{m_1 v_1 y^2}{2} - \frac{m_2 v_2}{2}$$

$$\times [y - b_1]^2 - \cdots - \frac{m_{N-1} v_{N-1}}{2} [y - b_{N-2}]^2 - D_y v_1 y - D_y a_1 \quad (4.31)$$

On substituting $y = b_p$ in Eq. (4.31) one obtains an equation valid at the $(p + 1)$th girder. Substituting for $P_1, P_2, \ldots, P_{N-1}$ and $m_1, m_2, \ldots, m_N$ as before and multiplying throughout by $6/S_p^3$, there results

$$\left\{ \left(\frac{b_p}{S_p} \right)^3 (1 + \lambda_1) - \left(\frac{b_p - b_1}{S_p} \right)^3 \lambda_1 - \frac{\eta_1}{2} \left(\frac{S_1}{S_p} \right)^3 \right\} k_1 a_1$$

$$+ \sum_{t=2}^{t=p} \left\{ - \left(\frac{b_p - b_{t-2}}{S_p} \right)^3 \left(\frac{c_{t-1}}{k_t} \right) + \left(\frac{b_p - b_{t-1}}{S_p} \right)^3 \right.$$

$$\times \left(1 + \frac{c_{t-1}}{k_t} + \lambda_t \right) - \left(\frac{b_p - b_t}{S_p} \right)^3 \lambda_t \right\} k_t a_t$$

$$+ \left\{ \frac{\eta_{p+1}}{2} \left(\frac{S_{p+1}}{S_p} \right)^3 - \lambda_p \frac{k_p}{k_{p+1}} \right\} k_{p+1} a_{p+1}$$

$$- \left\{ \frac{\eta_1}{2} \left(\frac{S_1}{S_p} \right)^2 \left(\frac{b_p}{S_p} \right) + 3\mu_1 \left(\frac{S_1}{S_p} \right) \left(\frac{b_p}{S_p} \right)^2 \right\} k_1 S_1 v_1$$

$$- \sum_{t=2}^{t=p} 3\mu_t \left(\frac{S_t}{S_p} \right) \left(\frac{b_p - b_{t-1}}{S_p} \right)^2 k_t S_t v_t$$

$$= P \left[\frac{b_p - b_e}{S_p} \right]^3 \quad (4.32)$$

Equation (4.32) is available for $p = 1, 2, \ldots, (N - 1)$. Once again the terms involving summation from 2 to p are absent when $p = 1$.

Summary of the steps in the solution. The complete set of equations to be solved is as follows:

1. The first equation is for vertical equilibrium of the whole system and is given as Eq. (4.25).

2. The second equation is for moment equilibrium of the whole system and is given as Eq. (4.27). For convenience, it is divided on both sides by b_{N-1}.

3. The next $(N - 1)$ equations are for slope conditions for girders 2 to N, it being noted that the slope condition for girder 1 is used up in the form of a constant of integration. These equations are derived from Eq. (4.30) by putting successively p equal to 1, 2, ..., and $(N - 1)$.

4. The remaining $(N - 1)$ equations are for deflection conditions at girders 2 to N, it being similarly noted that the deflection condition for girder 1 is used up in the form of a constant of integration. These equations are derived from Eq. (4.32) by putting successively p equal to 1, 2, ..., and $(N - 1)$.

For solution purposes, both sides of all these equations are divided by P, and fractions $\rho_1, \rho_2, \ldots, \rho_N$ and $\rho_1^*, \rho_2^*, \ldots, \rho_N^*$ are defined as below:

$$\rho_1 = \frac{k_1 a_1}{P}$$

$$\rho_2 = \frac{k_2 a_2}{P}$$

$$\vdots$$

$$\rho_N = \frac{k_N a_N}{P}$$

(4.33)

$$\rho_1^* = \frac{k_1 S_1 v_1}{P}$$

$$\rho_2^* = \frac{k_2 S_2 v_2}{P}$$

$$\vdots$$

$$\rho_N^* = \frac{k_N S_N v_N}{P}$$

(4.34)

Fractions $\rho_1, \rho_2, \ldots, \rho_N$ are referred to as distribution coefficients for longitudinal bending moments for girders 1, 2, ..., N respectively.

Similarly, $\rho_1^*, \rho_2^*, \ldots, \rho_N^*$ are distribution coefficients for longitudinal twisting moments in girders 1, 2, ..., N respectively.

Equally spaced girders. For a bridge with N equally spaced girders of equal stiffnesses, the 2N equations are as follows:

$$\rho_1 + \rho_2 + \cdots + \rho_N = 1.0$$

$$(N - 1 + \lambda)\rho_1 + (N - 2)\rho_2 + (N - 3)\rho_3 + \cdots + \rho_{N-1} - \lambda\rho_N$$
$$- \mu\rho_1^* - \mu\rho_2^* - \cdots - \mu\rho_N^* = (N - 1 - e)$$

For $r = 2, 3, \ldots, N,$

$$\{(r - 1)^2 + \lambda(2r - 3)\}\rho_1 + \sum_{t=2}^{t=r-1} \{(r - t)^2 - 2\lambda\}\rho_t - \lambda\rho_r$$

$$- \left\{2\mu(r - 1) + \frac{\eta}{6}\right\}\rho_1^* - 2\mu \sum_{t=2}^{t=r-1} (r - t)\rho_t^* + \frac{\eta}{6}\rho_r^* = [r - 1 - e]^2$$

For $r = 2, 3, \ldots, N,$

$$\left\{(r - 1)^3 + \lambda(3r^2 - 9r + 7) - \frac{\eta}{2}\right\}\rho_1 + \sum_{t=2}^{t=r-1}$$

$$\times \{(r - t)^3 - 6\lambda(r - t)\} - \left(\lambda - \frac{\eta}{2}\right)\rho_r - \left\{\frac{\eta}{2}(r - 1) + 3\mu(r - 1)^2\right\}$$

$$\times \rho_1^* - 3\mu \sum_{t=2}^{t=r-1} (r - t)^2\rho_t^* = [r - 1 - e]^3$$

$$(4.35)$$

It may be noted that the second equation of Eq. (4.35), for moment equilibrium, has been obtained by taking moments about the right-hand end, in contrast to Eq. (4.27). The two are clearly equivalent and differ from one another only by $(N - 1)$ times the vertical-equilibrium equation. It may sometimes be found more convenient in writing a computer program to use one form rather than the other.

In the above equations, the distance b_e of the external load from the left-hand-girder position is expressed as eS; the terms involving summation from $t = 2$ to $t = r - 1$ make a contribution only when r is 3 or more; the terms $[r - 1 - e]^2$ and $[r - 1 - e]^3$ are put equal to zero if their contents are negative, and the various constants, which change with every harmonic, are defined below for harmonic number m.

$$\eta = \frac{12}{(m\pi)^4}\left(\frac{L}{S}\right)^3 \frac{LD_y}{EI}$$

$$\lambda = \frac{1}{(m\pi)^2}\left(\frac{L}{S}\right)^2 \frac{SD_{yx}}{EI} \tag{4.36}$$

$$\mu = \frac{1}{(m\pi)^2}\left(\frac{L}{S}\right)^2 \frac{GJ}{EI}$$

The bank of equations defined by Eq. (4.35) can be written in the following matrix notation:

$$[A]\{\rho\} = \{R\} \tag{4.37}$$

General case.   For the general case, in which the girder spacings and stiffnesses are not equal, the bank of equations is of the following form:

$$
\begin{bmatrix} \text{Generalized} \\ A \text{ matrix} \end{bmatrix}
\begin{Bmatrix} \rho_1 \\ \rho_2 \\ \rho_3 \\ \cdot \\ \cdot \\ \rho_N \\ \rho_1^* \\ \cdot \\ \cdot \\ \rho_N^* \end{Bmatrix}
=
\begin{Bmatrix}
1 \\
b_e/b_{N-1} \\
\left[\dfrac{b_1 - b_e}{S_1}\right]^2 \\
\cdot \\
\cdot \\
\left[\dfrac{b_{N-1} - b_e}{S_{N-1}}\right]^2 \\
\left[\dfrac{b_1 - b_e}{S_1}\right]^3 \\
\cdot \\
\cdot \\
\left[\dfrac{b_{N-1} - b_e}{S_{N-1}}\right]^3
\end{Bmatrix}
\tag{4.38}
$$

Expressions for the various terms of the generalized $[A]$ matrix are given in App. III. As in the previous case, the values of parameters λ, η, and μ change with every harmonic number. For the general case their values for harmonic m are obtained by the following set of equations for $r = 1, 2, \ldots, (N - 1)$ in each case:

$$\eta_r = \frac{12}{(m\pi)^4} \left(\frac{L}{S_r}\right)^3 \frac{LD_y}{(EI)_r}$$

$$\lambda_r = \frac{1}{(m\pi)^2} \left(\frac{L}{S_r}\right)^2 \frac{S_r D_{yx}}{(EI)_r} \tag{4.39}$$

$$\mu_r = \frac{1}{(m\pi)^2} \left(\frac{L}{S_r}\right)^2 \frac{(GJ)_r}{(EI)_r}$$

Thus the $[A]$ matrix changes with every harmonic, and the $\{\rho\}$ vector should therefore be solved separately for each harmonic.

It is noted again that the bracketed terms of the $\{R\}$ vector should be put equal to zero if their contents are negative.

The $\{R\}$ vector can be calculated separately for each line of wheels, and the set of equations can be solved for each line of wheels. Alternatively, if two or more lines of wheels are identical, the various $\{R\}$ vectors for these different lines of wheels can be added together to give a single vector, using the set of equations which can be solved once for all lines of wheels simultaneously.

4.4   Calculation of Longitudinal Bending Moments

Bending moments in longitudinal girders are referred to as *longitudinal moments*. Determination of these moments is discussed in the following subsections.

Free moments.   The longitudinal-bending-moment diagram due to one longitudinal line of wheels, which gives the bending moments that a girder would carry if it sustained this load on its own, is here termed the *free moment;* at a distance x from the left-hand support this free moment is denoted by M_L. In a beam of span L and subjected to concentrated loads $W_1, W_2, \ldots, W_r$ at distances $x_1, x_2, \ldots, x_r$ respectively from the left-hand support, the free-bending-moment diagram is the familiar polygonal one. As shown in Chap. 2, this can be analyzed into a harmonic series as follows:

$$M_L = \frac{2L}{\pi^2} \sum_{p=1}^{p=r} W_p \left(\sin \frac{\pi x_p}{L} \sin \frac{\pi x}{L} + \frac{1}{2^2} \sin \frac{2\pi x_p}{L} \sin \frac{2\pi x}{L} \right.$$

$$\left. + \frac{1}{3^2} \sin \frac{3\pi x_p}{L} \sin \frac{3\pi x}{L} + \cdots \right) \tag{4.40}$$

where the notation is as shown in Fig. 4.15.

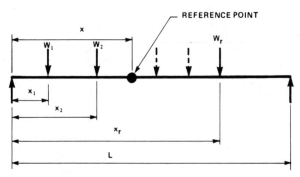

Figure 4.15 Series of point loads on a beam.

For convenience, Eq. (4.40) is rewritten as follows:

$$M_L = K_1 \sin \frac{\pi x}{L} + K_2 \sin \frac{2\pi x}{L} + K_3 \sin \frac{3\pi x}{L} + \cdots \quad (4.41)$$

where

$$K_1 = \frac{2L}{\pi^2} \sum_{p=1}^{p=r} W_p \sin \frac{\pi x_p}{L}$$

$$K_2 = \frac{2L}{\pi^2} \sum_{p=1}^{p=r} \frac{1}{2^2} W_p \sin \frac{2\pi x_p}{L}$$

$$\vdots$$

$$(4.42)$$

$$K_m = \frac{2L}{\pi^2} \sum_{p=1}^{p=r} \frac{1}{m^2} W_p \sin \frac{m\pi x_p}{L}$$

The magnitudes of constants $K_1, K_2, \ldots, K_m$ generally decrease as the harmonic number increases; in other words, K_m gets generally smaller as m gets larger because of the factor $1/m^2$. The physical significance of Eq. (4.41) can be demonstrated with the help of an example of a beam carrying five point loads, as shown in Fig. 4.16. In this figure, the actual beam moments are compared with those obtained by only the first term of the series represented by Eq. (4.41). It can be seen that the continuous curve of first harmonic moments clings very closely to the actual polygonal-moment diagram. It should be noted, however, that the comparison shown in Fig. 4.16 is given only to show the effectiveness of load representation by harmonic series and not to demonstrate the accuracy of the semicontinuum method. As shown later in this section, the semicontinuum method provides even more accurate results.

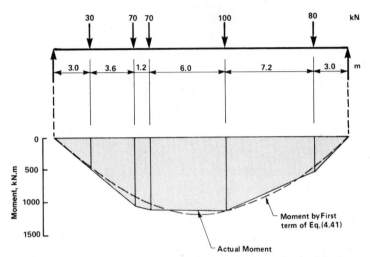

Figure 4.16 Comparison of actual moments with those obtained by harmonic analysis.

Distributed moments. Each term on the right-hand side of Eq. (4.42) represents beam moments which result from a harmonic loading corresponding to a harmonic number. For example, K_3 is the amplitude of the third harmonic of M_L and is derived from the third harmonics of all the loads. Having obtained the vector of distribution coefficients, i.e., $\{\rho\}$, for each harmonic in turn, the next step consists of applying the distribution coefficients to each term on the right-hand side of Eq. (4.41). By denoting the distribution coefficients for bending moments in girder n corresponding to harmonic m as $(\rho_n)_m$, the following expression is obtained from Eq. (4.41) for bending moment M_n in girder n at a distance x from the left-hand support:

$$M_n = (\rho_n)_1 K_1 \sin \frac{\pi x}{L} + (\rho_n)_2 K_2 \sin \frac{2\pi x}{L} + \cdots \qquad (4.43)$$

It has been shown in Chap. 2 that the convergence of free moment, as given by the right-hand side of Eq. (4.41), is relatively slow when there are only a few point loads on the beam. However, the convergence of moments obtained by Eq. (4.43) depends not only on the convergence of the series represented by Eq. (4.41) but also on successive values of $(\rho_n)_m$.

Avoidance of slow convergence. It is instructive to examine how the load effects due to harmonic effects are distributed. With the help of the numerical example given in Sec. 4.10, it is demonstrated that as the harmonic number increases, the load effects are increasingly re-

tained by the externally loaded girder. Analyses have confirmed that
in most cases virtually all effects of loading due to fourth and higher
harmonics are retained by the externally loaded girder. With reference
to Eq. (4.43), this means that if girder n is the externally loaded one,
then $(\rho_n)_4$ and all subsequent distribution coefficients are very close to
1.0, while if girder n is not the externally loaded one, then $(\rho_n)_4$ and
all subsequent distribution coefficients are very close to zero.

By taking advantage of this property of harmonic loads, the convergence of results can be hastened considerably by subtracting from the
free-response diagram those load effects which are distributed to other
longitudinal girders in order to obtain the bending-moment diagram
for the moments which are retained in the externally loaded girder.
For example, if the center girder of a three-girder bridge carries a
concentrated load at midspan, as shown in Fig. 4.17a, it is frequently
found in practical bridge designs that only first harmonic components
of the free-bending-moment diagram M_L are distributed to the outer
girders, with all higher harmonics being retained in the loaded girder.
In that event the bending moments in the loaded girder are very conveniently obtained as shown in Fig. 4.17b. This shows the free-bending-
moment diagram M_L, the first harmonic moments distributed to two
outer girders, and the bending moments accepted by the middle girder
as the difference between the two. The influence of higher harmonics
is extremely small. The total bending moment taken by the outer
girders if the first three harmonics are considered is shown schematically by the dashed line in Fig. 4.17b.

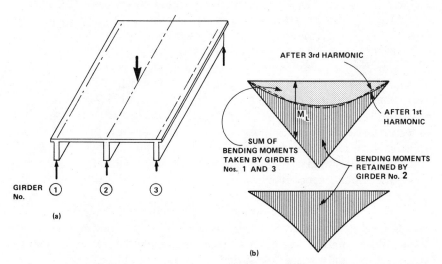

Figure 4.17 Distribution of moments in a three-girder bridge. (a) Bridge subjected to a
central-point load. (b) Determination of moments retained by the loaded girder.

The values of ρ for a very large harmonic number correspond to the case in which no transfer of load takes place between girders. For a girder n, this coefficient is denoted by $(\rho_n)_\infty$. When loads on only one longitudinal line are considered, then the sum of $(\rho_n)_\infty$ for all girders must be equal to 1.0. In the case of several longitudinal lines of similar loads, the sum of $(\rho_n)_\infty$ for all girders is equal to the number of longitudinal lines of loads.

From Eqs. (4.41) and (4.43) it can be readily shown that the distribution moment M_n for girder n is given by the following series:

$$M_n = (\rho_n)_\infty M_L - K_1[(\rho_n)_\infty - (\rho_n)_1] \sin \frac{\pi x}{L}$$

$$- K_2[(\rho_n)_\infty - (\rho_n)_2] \sin \frac{2\pi x}{L} - K_3[(\rho_n)_\infty - (\rho_n)_3] \sin \frac{3\pi x}{L} - \cdots \quad (4.44)$$

As the harmonic number m increases, the value of $(\rho_n)_m$ quickly approaches the value $(\rho_n)_\infty$, thus making the successive terms of the series very small and the convergence very rapid. The convergence of longitudinal moments as obtained by Eqs. (4.43) and (4.44) is discussed in the example of Sec. 4.10; it is shown there that results obtained by Eq. (4.44) converge very close to the true solution after distributing only the first harmonic, i.e., if one ignores completely the terms involving $\sin (2\pi x/L)$, $\sin (3\pi x/L)$, etc. On the other hand, results obtained from Eq. (4.43) do not converge so closely even after 30 harmonics.

According to Eq. (4.43) moments are obtained by successively adding the various harmonic effects, and according to Eq. (4.44) they are obtained by subtracting the distributed harmonic effects from the free-response diagram. The same procedures are available for longitudinal shears, as is discussed below in Sec. 4.5. The effectiveness of these two procedures in achieving convergence is discussed in Ref. 7 with the help of a specific example. One set of results from the reference is reproduced in Fig. 4.18, which refers to a bridge with five girders carrying two eccentrically placed lines of wheels. The figure shows for different harmonics the moments at midspan and the shears at three-eighths span in a girder near the applied loads, as obtained by the above-mentioned procedures. It can be seen that the procedure represented by Eq. (4.44) leads to very quick convergence for both moments and shears. When the procedure represented by Eq. (4.43), that of adding harmonic effects, is used, moments do begin to converge after about 30 harmonics, but even at that high harmonic number shears are still far from having converged.

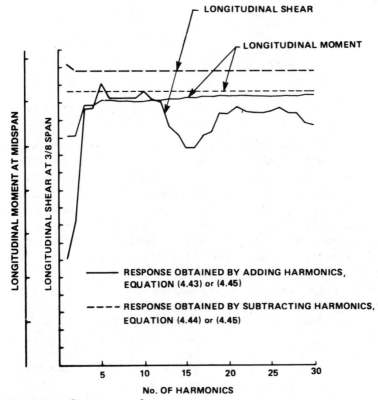

Figure 4.18 Convergence of responses.

4.5 Calculation of Longitudinal Shears

Vertical shear in a longitudinal girder is referred to as *longitudinal shear;* determination of this response is discussed in the following paragraphs.

Having obtained the values of the distribution coefficients ρ_1, $\rho_2, \ldots, \rho_n$, the longitudinal shears V_n retained by girder n can be obtained by differentiating either Eq. (4.43) or Eq. (4.44) with respect to the longitudinal coordinate x. This results in the following two equations:

$$V_n = (\rho_n)_1 K_1 \frac{\pi}{L} \cos \frac{\pi x}{L} + (\rho_n)_2 K_2 \frac{2\pi}{L} \cos \frac{2\pi x}{L} + \cdots \quad (4.45)$$

or
$$V_n = (\rho_n)_\infty V_L - K_1 \frac{\pi}{L} [(\rho_n)_\infty - (\rho_n)_1] \cos \frac{\pi x}{L}$$

$$- K_2 \frac{2\pi}{L} [(\rho_n)_\infty - (\rho_n)_2] \cos \frac{2\pi x}{L}$$

$$- K_3 \frac{3\pi}{L} [(\rho_n)_\infty - (\rho_n)_3] \cos \frac{3\pi x}{L} + \cdots \quad (4.46)$$

where V_L is the free shear due to one line of wheels, i.e., the shear obtained by treating a girder as an isolated beam. The distribution coefficients ρ are referred to in Sec. 4.3 as the coefficients for longitudinal bending moments. The use of the same coefficients for longitudinal shear does not imply that the transverse distribution pattern for longitudinal shear is the same as that for longitudinal moments. As shown in the numerical example of Sec. 4.10, the distribution patterns of longitudinal moments and shears obtained by the same coefficients (but different equations) can differ markedly from each other.

As in the case of moments by Eq. (4.43), the longitudinal shears by Eq. (4.45) are obtained by adding the effects due to each successive harmonic. Beam shears, as shown in Chap. 2, are quite slow to converge. Consequently, if Eq. (4.45) is used, a very large number of harmonics (say, over 50) will be required to attain good convergence.

The slow convergence can again be avoided by using Eq. (4.46) instead of Eq. (4.45). Once again advantage can be taken of the fact that

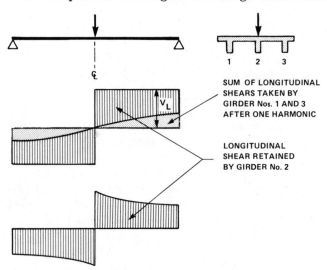

SUM OF LONGITUDINAL
SHEARS TAKEN BY
GIRDER Nos. 1 AND 3
AFTER ONE HARMONIC

LONGITUDINAL
SHEAR RETAINED
BY GIRDER No. 2

Figure 4.19 Distribution of longitudinal shears in a three-girder bridge.

higher harmonic effects in real-life bridges are retained virtually completely by the loaded girders.

The physical significance of Eq. (4.46) is illustrated in Fig. 4.19, which shows longitudinal shears in a three-girder bridge due to a point load at the midspan of the middle girder. In considering only the first harmonic, the shape of the shears distributed to the two outer girders is that of a cosine curve. Shears retained by the loaded girder are obtained by taking away from the free-shear diagram those shears which are distributed to the other girders. It is interesting to note that in the immediate vicinity of the load virtually no shear is distributed to the other girders. However, the portion of shears distributed to other girders increases as one moves away from the load. The consideration of second and higher harmonics changes the pattern only slightly. Hence, it can be concluded that the pattern of transverse distribution of shears is not independent of the longitudinal position of the reference section, as is assumed in the so-called distribution coefficient methods such as those given in Refs. 2 and 8.

4.6 Calculation of Transverse Bending Moments

Bending moments in the direction perpendicular to the flow of traffic are referred to as *transverse moments*. For determining transverse moments in slab-on-girder bridges, it is usual to divide this response into two parts known as *global moments* and *local moments* respectively. As shown in Fig. 4.20, the former response is obtained by ignoring the local bending of the deck slab between adjacent girders and the latter by assuming the girders to be unyielding. The semicontinuum method can accurately predict global transverse moments in linear elastic slab-on-girder bridges; the deformation of the cross section of such a bridge under global moments is shown in Fig. 4.20b. An approximate method of deriving local transverse moments by the semicontinuum method is given in this section and a more accurate method in Ref. 1. Attention is, however, drawn to the fact that, as shown in Ref. 1, the superposition of global and local transverse moments to obtain the total response is not strictly defensible but that it always leads to somewhat conservative (i.e., safe) results.

If preferred, the semicontinuum method can be used to derive transverse moments in their entirety without having to resort to the concept of global and local moments. If this application is taken, then it is necessary only to utilize Eq. (4.28) and thereby obtain the transverse moment intensities harmonic by harmonic and subsequently super-

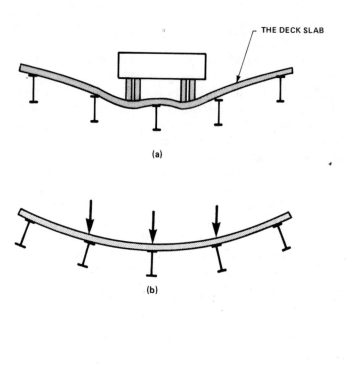

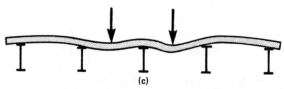

Figure 4.20 Cross-section deformations due to transverse moments. (*a*) Deformations due to overall transverse moments. (*b*) Deformations due to global transverse moments. (*c*) Deformations due to local transverse moments.

impose the results. A word of caution is necessary: the fiction of "point" load can lead to very high local intensities of transverse bending moment in the vicinity of the point load. In order to avoid this unrealistic local effect, one can proceed, at choice, in one of two ways. The first way is to retain the concept of the point load because of the convenience of its application in the design calculations but to disregard those harmonics in the harmonic representation of the point load whose half wavelength is less than the width of the actual applied concentrated load. The second way is to replace the point load immediately by a uniformly distributed load of the same total size and distribute it over a length equal to that of the actual applied load. If this second way is

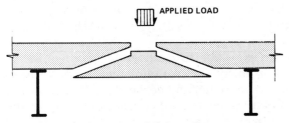

Figure 4.21 Punching-shear-failure mode.

adopted, the very high fictitious values of local transverse moment intensities will automatically be avoided.

Arching effect. Attention is also drawn to an arching action in deck slabs, as a result of which a deck slab, subjected to concentrated loads, fails by punching shear instead of by flexure. A typical model of such failure is shown in Fig. 4.21. The change of the failure mode from flexure to punching shear as a result of arching action, as shown in Fig. 4.22a, develops only when the material of the deck slab is weaker in tension then in compression. When the material of the deck slab has similar stress-strain characteristics in compression and tension, then, as shown in Fig. 4.22b, this arching action does not develop.

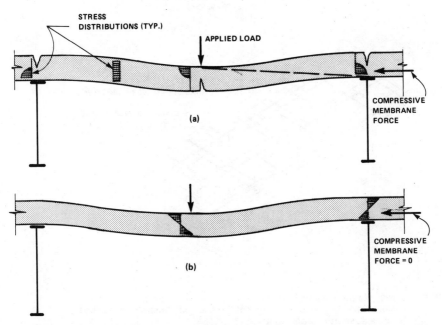

Figure 4.22 Compressive membrane action. (a) Deck slab which can crack. (b) Deck slab with the same stress-strain relationships in compression and tension.

A method has been developed in Ref. 4 to take account of the beneficial action of this arching effect. Details of this method are also reproduced in Ref. 1.

The arching effect method results in considerable savings in the amount of deck-slab reinforcement. Typical reinforcement patterns in a particular deck slab by conventional and new methods are shown in Fig. 4.23. It can be seen that the latter method leads to a simpler and cleaner reinforcement arrangement.

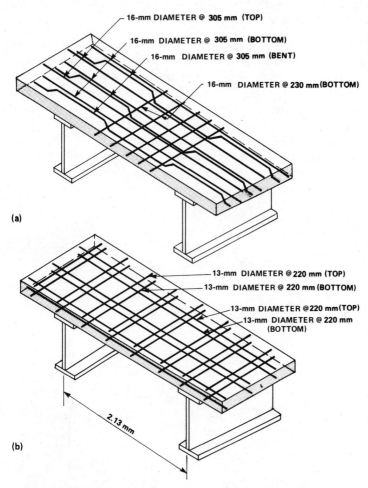

Figure 4.23   Reinforcement patterns in a deck slab. (*a*) Reinforcement by conventional design. (*b*) Reinforcement by the punching-shear method which takes account of internal arching.

When the method that takes account of internal arching in deck slabs is used, it is no longer necessary to calculate and design for the transverse moments. However, since only a very few codes permit this method to be used, transverse moments which ignore the arching action will continue to be determined. Accordingly, the analytical methods for determining this response in linear elastic slab-on-girder bridges, such as that shown in Fig. 4.22b, are given below.

Global transverse moments. If the designer chooses to utilize the concept of global and local moments, then the global transverse bending moments M_y follow directly from the distribution coefficients. By referring to Fig. 4.13 and Eq. (4.28), M_y can be obtained by expressing the support forces $P_1, P_2, \ldots, P_N$ and restraint moments $m_1 v_1$, $m_2 v_2, \ldots, m_N v_N$ in terms of the distribution coefficients.

The expression of the support forces $P_1, P_2, \ldots, P_N$ in terms of the distribution coefficients uses Eqs. (4.21), (4.22), (4.23), and (4.33) and gives the following:

$$P_1 = k_1[(1 + \lambda_1)a_1 - \lambda_1 a_2] = P\left[(1 + \lambda_1)\rho_1 - \frac{k_1}{k_2}\lambda_1\rho_2\right] \quad (4.47)$$

For $r = 2, 3, \ldots, (N - 1)$,

$$P_r = k_r\left[-\frac{c_{r-1}}{k_r}a_{r-1} + \left(1 + \frac{c_{r-1}}{k_r} + \lambda_r\right)a_r - \lambda_r a_{r+1}\right]$$

$$= P\left[-\lambda_{r-1}\rho_{r-1} + \left(1 + \frac{c_{r-1}}{k_r} + \lambda_r\right)\rho_r - \frac{k_r}{k_{r+1}}\lambda_r\rho_{r+1}\right] \quad (4.48)$$

$$P_N = k_N\left[-\frac{c_{N-1}}{k_N}a_{N-1} + \left(1 + \frac{c_{N-1}}{k_N}\right)a_N\right]$$

$$= P\left[-\lambda_{N-1}\rho_{N-1} + \left(1 + \frac{c_{N-1}}{k_N}\right)\rho_N\right] \quad (4.49)$$

By using Eq. (4.34), restraint moment at the location of girder r is given by

$$m_r v_r = P\frac{m_r}{k_r S_r}\rho_r^*$$

$$= \mu_r S_r P\rho_r^* \quad (4.50)$$

where r varies from 1 to N. For a given harmonic q, the quantity P in Eqs. (4.47), (4.48), (4.49), and (4.50) is given by

$$P_q = K_q \frac{q^2\pi^2}{L^2} \qquad (4.51)$$

where K_q is obtained from Eq. (4.42).

The Poisson's-ratio effect on transverse moments in slab-on-girder bridges is negligible. Therefore, the second term of Eq. (1.11) can be ignored, and M_y is given by

$$M_y = -D_y \frac{\partial^2 w}{\partial y^2} \qquad (4.52)$$

It is recalled from Eq. (4.28) that for a unit length of the transverse medium the first harmonic effects are given by

$$-D_y \frac{d^2 w}{dy^2} = \{P_1 y + P_2[y - b_1] + \cdots$$

$$+ P_{N-1}[y - b_{N-2}] - P[y - b_e]$$

$$- m_1 v_1 - [m_2 v_2] - \cdots - [m_{N-1} v_{N-1}]\} \quad (4.53)$$

Similar expressions are, of course, available for the higher harmonics, where, for the harmonic number under consideration, $P_1, P_2, \ldots,$ P_N can be obtained from Eqs. (4.47) and (4.48) and $m_1 v_1, m_2 v_2, \ldots,$ $m_N v_N$ from Eq. (4.50).

In Eq. (4.53) the term $[m_2 v_2]$ is ignored if $y < b_1$, the term $[m_2 v_3]$ is ignored if $y < b_2$, and so on. By considering the effects of all harmonics, it is readily seen that at a distance x from the left-hand supports the global transverse moment M_y is given by the following series:

$$M_y = \sum_{q=1}^{q=\infty} \left\{ (P_1)_q \sin \frac{q\pi x}{L} y + (P_2)_q \sin \frac{q\pi x}{L} [y - b_1] + \cdots \right.$$

$$+ (P_{N-1})_q \sin \frac{q\pi x}{L} [y - b_{N-2}] - (P)_q \sin \frac{q\pi x}{L} [y - b_e]$$

$$- (m_1 v_1)_q \sin \frac{q\pi x}{L} - [m_2 v_2]_q \sin \frac{q\pi x}{L} - \cdots$$

$$\left. - [m_{N-1} v_{N-1}]_q \sin \frac{q\pi x}{L} \right\} \qquad (4.54)$$

where the subscript q outside the brackets refers to the value of the term for harmonic q. The criteria for ignoring the bracketed terms remain the same as given immediately after Eq. (4.53).

Local moments by the semicontinuum method. When a concentrated load is situated on the deck slab between two adjacent girders, as shown in Fig. 4.24, the local transverse moments are obtained by assuming that the girders do not deflect. An estimate of these local moments can be obtained by the semicontinuum method in which the deck slab is idealized as an assembly of one-dimensional (1-D) beams and a continuous medium. Such an idealization for slab bridges is discussed in detail in Chap. 7.

A safe-side estimate of local transverse moments can readily be obtained if the following simplifying assumptions are made about the semicontinuum idealization.

1. The 1-D beams and the continuous medium are assumed to be torsionless. It is already established that such an assembly has worse distribution characteristics than its counterpart with nonzero torsional rigidities. Therefore, the results of analysis using the former idealization are necessarily conservative.

2. The deck slab is assumed to be fixed against rotation at the girder axes, so that for central loads the 1-D beams can be assumed to be simply supported at the inflexion points between the girders. For a girder spacing of S, these inflexion points are a distance $S/2$ apart. Again, this assumption leads to overestimation of load sustained by the loaded beam and is therefore conservative.

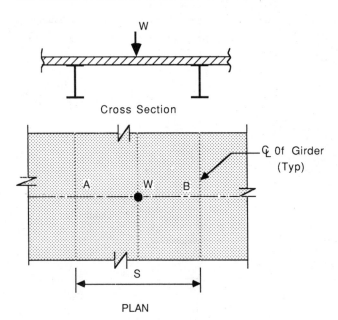

Figure 4.24 Concentrated load acting on a deck slab between girders.

Figure 4.25 shows four different idealizations for the deck slab in the immediate vicinity of the central load between two adjacent girders. In these idealizations, the deck slab is represented by three, four, and five strips of different widths. Since the idealizations are torsionless, both λ and μ are equal to zero. According to the theory developed in Chap. 3, the distribution factors corresponding to a symmetric central

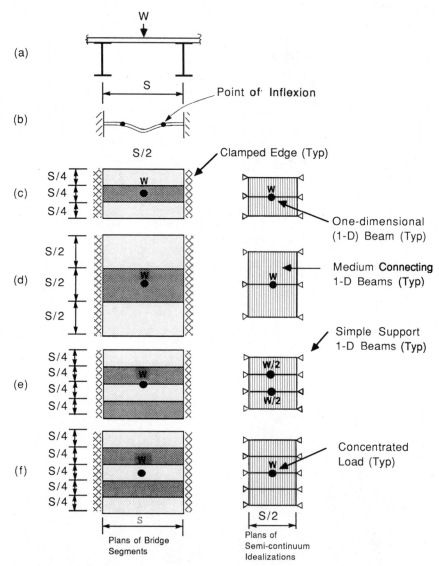

Figure 4.25 Idealization for analysis of local transverse moment. (*a*) Bridge cross section. (*b*) Idealized cross section. (*c*)–(*f*) Idealizations by considering different strips.

load on assemblies with three, four, and five main beams, respectively, are as follows.

1. For a three-beam assembly shown in Fig. 4.25c and d,

$$\rho_1 = \frac{\eta}{4 + 3\eta}$$

$$\rho_2 = \frac{4 + \eta}{4 + 3\eta} \tag{4.55}$$

$$\rho_3 = \rho_1$$

2. For a four-beam assembly shown in Fig. 4.25e,

$$\rho_1 = \frac{2\eta - 3}{8\eta + 40}$$

$$\rho_2 = \frac{2\eta + 23}{8\eta + 40} \tag{4.56}$$

$$\rho_3 = \rho_2$$

$$\rho_4 = \rho_1$$

3. For a five-beam assembly shown in Fig. 4.25f,

$$\rho_1 = \frac{\eta(\eta - 6)}{5\eta^2 + 68\eta + 28} \qquad \rho_2 = \frac{\eta(\eta + 22)}{5\eta^2 + 68\eta + 28}$$

$$\rho_3 = \frac{\eta^2 + 36\eta + 28}{5\eta^2 + 68\eta + 28} \qquad \rho_4 = \rho_2 \tag{4.57}$$

$$\rho_5 = \rho_1$$

For a solid slab D_x is equal to D_y. Therefore, the expression for η for a solid slab in two-way action is given by

$$\eta = \frac{12}{\pi^4} \left(\frac{L}{S_g} \right)^4 \tag{4.58}$$

where L is the distance between points of inflexion, i.e., $0.5S$, and S_g is the spacing of the 1-D beams.

Values of η and the various distribution coefficients for the four idealizations shown in Fig. 4.25 are listed in Table 4.1.

It is assumed that the distribution coefficients represent the distribution of negative moments in the deck slab over girders. Thus, a

TABLE 4.1 Distribution Coefficients for Semicontinuum Idealizations Shown in Fig. 4.25

Idealization shown in figure no.	η	Values of distribution coefficient				
		ρ_1	ρ_2	ρ_3	ρ_4	ρ_5
4.25c	1.97	0.200	0.600	0.200		
4.25d	0.12	0.280	0.944	0.028		
4.25e	1.97	0.021	0.479	0.479	0.021	
4.25f	1.97	−0.043	0.261	0.565	0.261	−0.043

distribution coefficient for a 1-D beam is equal to the fraction of the negative moment taken by the strip of the deck slab represented by the beam. The intensity of restraint moment M_y taken by any strip would be equal to the moment taken by the strip divided by the strip width. Therefore,

$$M_y = \frac{\rho M_T}{S_g} \tag{4.59}$$

where ρ is the relevant distribution coefficient, M_T is the total negative moment over a girder, and S_g is the spacing of 1-D beams. The quantity M_y/S is plotted as histograms for the four idealizations in Fig. 4.26.

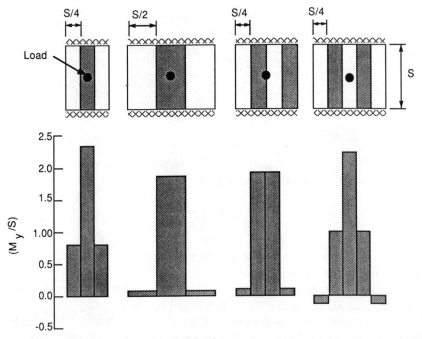

Figure 4.26 Histograms for transverse local moments.

The physical significance of these histograms can be explained with reference to one particular idealization, say, the one with three strips each of width $S/4$. In this case, the middle strip of width $S/4$, which is symmetrically situated about the point load, will sustain 60 percent of the total restraint moment. The two adjacent strips of width $S/4$ will take 20 percent each. The accuracy of analysis is, of course, related to the selected idealization. As can be seen in Fig. 4.26, the four idealizations lead to apparently different results. The differences, however, diminish significantly when a continuum distribution of the restraint moments is sketched over the histograms. While a detailed account of this technique is given in Chap. 7, the process is illustrated in Fig. 4.27. The continuum curve is plotted in such a way that the area under it for any strip is very closely equal to the corresponding area of the histogram. Of course, the total area under a histogram, which is equal to the total restraint moment in the slab, should also be equal to the area under the corresponding curve.

The continuum restraint-moment intensity curves are plotted in Fig. 4.28 for the four idealizations shown in Fig. 4.26. These curves are also superimposed upon each other in Fig. 4.28e, from which it is clear that the differences between the various curves are not as large as those of

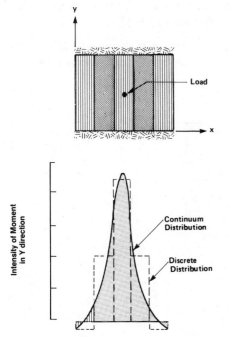

Figure 4.27 Plotting of continuum distribution of moments in solid slabs.

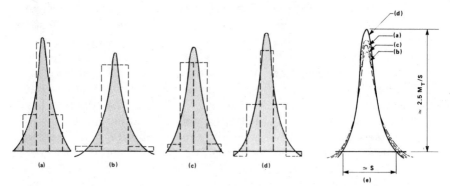

Figure 4.28 Continuum moment intensity curves for the four idealizations shown in Fig. 4.26.

the histograms shown in Fig. 4.26. The following conclusion can be drawn from Fig. 4.28 for obtaining local transverse moments in deck slabs. These conclusions apply equally to sagging moments between girders and to hogging moments over girders.

1. Virtually all the moment is contained within a width S measured along the girder.

2. The maximum intensity of local moment is $2.5/S$ of the total moment M_T. In other words, the total transverse local moment can be safely assumed to be taken by a strip of width $0.4S$.

The analysis given above lends direct support to the idea of *dispersion* of the concentrated load as a device for design. A dispersion angle on

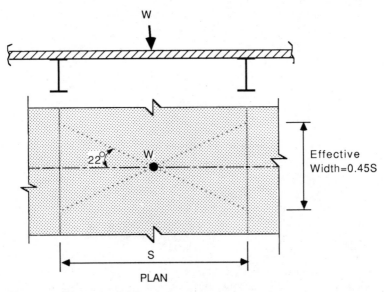

Figure 4.29 Dispersion angle.

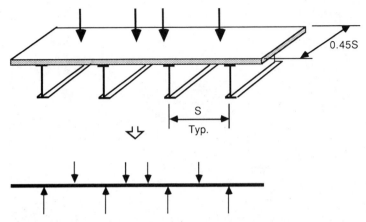

Figure 4.30 Approximate way of determining local transverse moments.

either side of the transverse line through the load can be taken to identify the width of slab along the girder, which can be assumed to sustain all the load. As shown in Fig. 4.29, the angle of dispersion derived from the above criteria is found to be about 22°.

For loads along a transverse line, the local transverse moments can be approximately determined by analyzing a transverse slice of a bridge of width $0.45S$ as a continuous beam. This process is diagrammatically shown in Fig. 4.30. It should be noted that for this analysis to be valid adjacent transverse lines of wheels should be at least a distance S apart.

4.7 Calculation of Transverse Shear

Vertical shear in a plane passing through a longitudinal line in the deck slab is referred to here as the *transverse shear*. This response is not of consequence in slab-on-girder bridges. However, when transverse flexural rigidity is small, as it is in multibeam bridges, load transfer between longitudinal beams takes place mainly through transverse shear. Accordingly, for such bridges this response must be calculated. It is noted that the design of shear keys in multibeam bridges is governed by transverse shear.

The method of calculating transverse shear given here is valid only when the loading is applied directly over girders and not between them. Such a restriction, however, is not unreasonable for multibeam bridges. In such bridges, as shown in Fig. 4.31, the widths of joints containing the shear keys are small as compared with the beam spacings. It can be readily demonstrated that for such cases the value of transverse shear in a joint is little affected by the transverse position of loads on the adjacent beams. Transverse shear V_y at a distance x from the left-

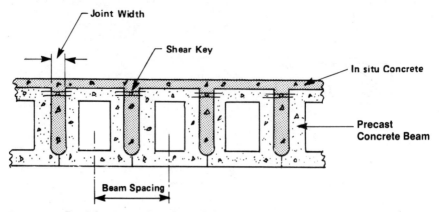

Figure 4.31 Partial cross section of a multibeam bridge.

hand support and a distance y from the left-hand girder can be obtained by differentiating the right-hand side of Eq. (4.54) with respect to y. This gives

$$V_y = \sum_{q=1}^{\infty} \left\{ (P_1)_q \sin \frac{q\pi x}{L} + \left[(P_2)_q \sin \frac{q\pi x}{L} \right]_{b_1} + \cdots \right.$$
$$\left. + \left[(P_{N-1})_q \sin \frac{q\pi x}{L} \right]_{b_{N-1}} - \left[(P)_q \sin \frac{q\pi x}{L} \right]_{b_e} \right\} \quad (4.60)$$

where in the above the term $[(P_2)_q \sin (q\pi x/L)]_{b_1}$ is ignored if $y < b_1$, the term $[(P_3)_q \sin (q\pi x/L)]_{b_2}$ is ignored if $y < b_2$, and so on. The subscript q outside the parentheses refers to the value of the term for harmonic q.

It can be appreciated that transverse shear depends upon the load passed on from a loaded girder to adjacent ones. As explained in Sec. 4.4, loads represented by higher harmonics are almost entirely retained by the loaded girders. It follows, therefore, that transverse shear due to higher harmonics will be extremely small. Because of this phenomenon, the summation represented by Eq. (4.60) converges very quickly, with virtually complete convergence obtained by as few as three harmonics (i.e., taking $q = 1, 2,$ and 3 only).

4.8 Calculation of Longitudinal Twisting Moments

As shown in Fig. 4.9, the applied twisting-moment intensity corresponding to the first harmonic loading is in the shape of a half sine wave. As shown in Sec. 4.2, for girder r this intensity of applied twisting moment t_x is given by

$$t_x = (GJ)_r \frac{\pi^2}{L^2} v_r \sin \frac{\pi x}{L} \qquad (4.61)$$

The girder torque is obtained by integrating the intensity of applied twisting moments with respect to x. Thus, by integrating the right-hand side of Eq. (4.61), the torque T_r in girder r for the first harmonic is given by

$$(T_r)_1 = (GJ)_r \frac{\pi}{L} (v_r)_1 \cos \frac{\pi x}{L} \qquad (4.62)$$

where the subscripts r to T and v refer to the harmonic numbers. The shapes of distributions of applied twisting-moment intensity and girder torque along a girder are shown in Fig. 4.32.

For harmonic m, the torque $(T_r)_n$ is given by

$$(T_r)_m = (GJ)_r \frac{m\pi}{L} (v_r)_m \cos \frac{m\pi x}{L} \qquad (4.63)$$

Dealing again with the first harmonic, the distribution factor for twisting moments in girder r, i.e., ρ_r^*, is given by Eq. (4.34):

$$(\rho_r^*)_1 = \frac{k_r S_r v_r}{P} \qquad (4.64)$$

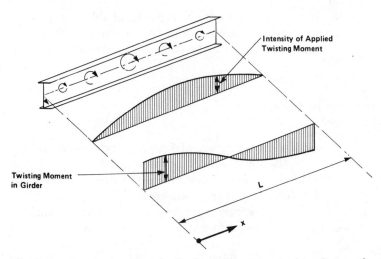

Intensity of Applied
Twisting Moment

Twisting Moment
in Girder

L

x

Figure 4.32 Distribution of longitudinal twisting moments along the span for the first harmonic.

On eliminating the amplitude v_r between Eqs. (4.62) and (4.64) and using the definition of μ_r and k_r given in Eq. (4.19),

$$(T_r)_1 = \left[(\mu_r)_1 \right) \frac{\pi S_r}{L} (\rho_r^*)_1 \right] \frac{PL^2}{\pi^2} \cos \frac{\pi x}{L} \tag{4.65}$$

Since PL^2/π^2 is the amplitude of the first harmonic of the free-bending-moment diagram, identified as K_1, in Eq. (4.41), Eq. (4.65) may be written as

$$(T_r)_1 = (f_r)_1 K_1 \cos \frac{\pi x}{L} \tag{4.66}$$

where
$$(f_r)_1 = (\mu_r)_1 \frac{\pi S_r}{L} (\rho_r^*)_1 \tag{4.67}$$

The total torque T_r in girder r is given by summing the effects of all the harmonics. Thus,

$$T_r = (f_r)_1 K_1 \cos \frac{\pi x}{L} + (f_r)_2 K_2 \cos \frac{2\pi x}{L} + \cdots \tag{4.68}$$

where, for harmonic m, $(f_r)_m$ is given by

$$(f_r)_m = (\mu_r)_m \frac{m\pi S_r}{L} (\rho_r^*)_m \tag{4.69}$$

with $(\mu_r)_m$ given by Eq. (4.39).

In practice the successive terms of the right-hand side of Eq. (4.68) rapidly become small. As demonstrated in the example in Sec. 4.10, convergence with good accuracy can be obtained by the first two harmonics.

4.9   Calculation of Transverse Twisting Moments

The transverse twisting moments act about transverse axes and are sustained entirely by the deck slab. It is assumed that at a given longitudinal position this response has constant values between two adjacent girders. The intensity of the transverse twisting moment between girders r and $(r + 1)$ and corresponding to harmonic m is denoted by $(\tau_r)_m$. In Sec. 4.2, it has been shown that $(\tau_1)_1$ is given by

$$(\tau_1)_1 = \frac{D_{yx}}{S_1} \frac{\pi}{L} (a_2 - a_1) \cos \frac{\pi x}{L} \tag{4.70}$$

Thus, as shown in Fig. 4.33, the transverse twisting moment varies along the span in the shape of a half cosine wave. By substituting for a_1 and a_2 in terms of k_1, k_2, and P from Eq. (4.33), Eq. (4.70) can be rewritten as

$$(\tau_1)_1 = \frac{D_{yx}}{S_1} \frac{\pi}{L} \rho \left(\frac{\rho_2}{k_2} - \frac{\rho_1}{k_1} \right) \cos \frac{\pi x}{L} \qquad (4.71)$$

Once again it is convenient to work in terms of K_1, which is equal to PL/π^2. It is also convenient to substitute k_1 and k_2 by using Eq. (4.16). The result of these substitutions is

$$(\tau_1)_1 = \frac{\pi K_1}{S_1} (S_2 \lambda_2 \rho_2 - S_1 \lambda_1 \rho_1) \cos \frac{\pi x}{L} \qquad (4.72)$$

For the general case, the twisting moment τ_r due to all harmonics is given by

$$\tau_r = \sum_{m=1,2,\ldots}^{\infty} \frac{(m\pi)K_m}{S_r} (S_{r+1}\lambda_{r+1}\rho_{r+1} - S_r \lambda_r \rho_r) \cos \frac{m\pi x}{L} \qquad (4.73)$$

Equation (4.73) is valid for $r = 1, 2, \ldots, (n-1)$, and for each value of m in turn λ_r and λ_{r+1} take the values as defined by Eq. (4.39).

τ_r also converges very rapidly. As shown in Sec. 4.10, the consideration of only the first two harmonics is sufficient to obtain convergence with good accuracy.

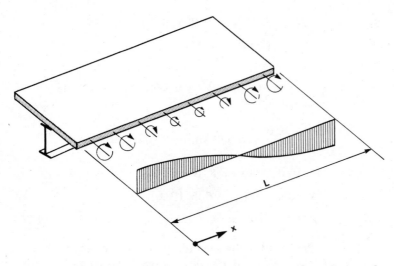

Figure 4.33 Distribution of transverse twisting moments along the span for the first harmonic.

4.10 Worked Example

The application of the semicontinuum method of analysis is illustrated here with the help of a specific example which considers a bridge with five equally spaced girders having the following relevant details.

Span, L	1016
Girder spacing, S	99
Slab thickness	7.5
Modulus of elasticity, E, of girder and slab material	3×10^6
Shear modulus of girder and slab material	1.5×10^6
Moment of inertia, I, of each girder	6.124×10^5
Torsional inertia, J, of each girder	0.177×10^5

The bridge, as shown in Fig. 4.34, is subjected to two lines of loads each consisting of three loads. Details of loads and their positions are also given in Fig. 4.34. The absence of units for the numerical quantities given above and in Fig. 4.34 is deliberate and is for the sake of generality. All the above quantities are of course given in a consistent set of units.

Calculation of rigidities. The values of the moment of inertia and torsional inertia of a girder given above were obtained by using standard procedures, which, for example, are given in Ref. 1. According to these procedures, D_y and D_{yx} are given by

$$D_y = Et^3/12 = 1.054 \times 10^8$$
$$D_{yx} = Gt^3/6 = 1.054 \times 10^8$$

(4.74)

Free-moment diagram. The coefficient K_1 is obtained by Eq. (4.42) as follows for one line of wheels:

$$K_1 = \frac{2 \times 1016}{\pi^2} [16,000 \sin (\pi \times 381/1016)$$

$$+ 16,000 \sin (\pi \times 508/1016)$$

$$+ 4000 \sin (\pi \times 635/1016)]$$

$$= 7.100 \times 10^6$$

(4.75)

Similarly, the coefficients for the second, third, fourth, and fifth harmonics, namely, K_2, K_3, K_4, and K_5, are found to be (0.437×10^6), (-0.541×10^6), (-0.154×10^6), and (0.069×10^6) respectively. Thus,

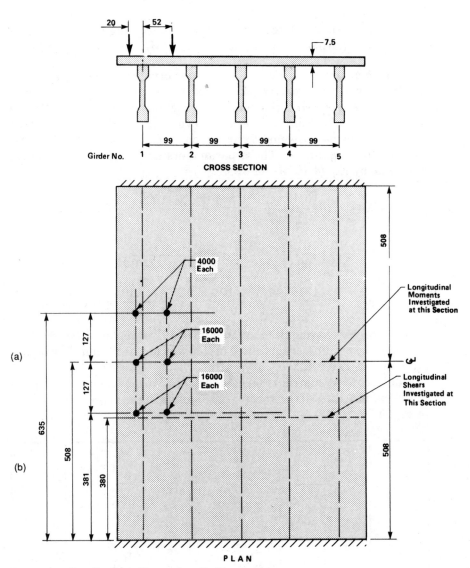

Figure 4.34 Details of a bridge and applied loading.

Eq. (4.40) for the three loads in one line can be rewritten as follows:

$$M_L = 7.100 \times 10^6 \sin \frac{\pi x}{L} + 0.437 \times 10^6 \sin \frac{2\pi x}{L}$$

$$- 0.541 \times 10^6 \sin \frac{3\pi x}{L} - 0.154 \times 10^6 \sin \frac{4\pi x}{L} \quad (4.76)$$

As is demonstrated later, it is sufficiently accurate to consider only the first four harmonics in the equation for M_L.

Calculation of matrix [A].   For a five-girder bridge with equally spaced similar girders, the matrix [A], which is defined by Eqs. (4.37) and (4.38), is developed with the help of instructions given in App. III. The resulting [A] matrix has the terms shown in Fig. 4.35. It is noted that the various terms of the matrix are given in terms of the dimensionless parameters λ, η, and μ. These parameters are obtained for the first harmonic by Eq. (4.19) as follows:

$$\lambda = \frac{1}{\pi^2}\left(\frac{1016}{99}\right)^2 \frac{99 \times 1.054 \times 10^8}{3 \times 10^6 \times 6.124 \times 10^5} = 0.0606$$

$$\eta = \frac{12}{\pi^4}\left(\frac{1016}{99}\right)^3 \frac{1016 \times 1.054 \times 10^8}{3 \times 10^6 \times 6.124 \times 10^5} = 7.766 \qquad (4.77a)$$

$$\mu = \frac{1}{\pi^2}\left(\frac{1016}{99}\right)^2 \frac{1.5 \times 10^6 \times 0.177 \times 10^5}{3 \times 10^6 \times 6.124 \times 10^5} = 0.154$$

For a higher harmonic n, π in Eq. (4.19) is replaced by $n\pi$. The values of the parameters obtained in this way for the various harmonics are listed in Table 4.2.

A different [A] matrix is formed for each harmonic by substituting the relevant values of the various parameters in the expressions shown in Fig. 4.35.

1	1	1	1	1	0	0	0	0	0
$4+\lambda$	3	2	1	$-\lambda$	$-\mu$	$-\mu$	$-\mu$	$-\mu$	$-\mu$
$1+\lambda$	$-\lambda$	0	0	0	$-\left(\dfrac{\eta}{6}+2\mu\right)$	$\dfrac{\eta}{6}$	0	0	0
$4+3\lambda$	$1-2\lambda$	$-\lambda$	0	0	$-\left(\dfrac{\eta}{6}+4\mu\right)$	-2μ	$\dfrac{\eta}{6}$	0	0
$9+5\lambda$	$4-2\lambda$	$1-2\lambda$	$-\lambda$	0	$-\left(\dfrac{\eta}{6}+6\mu\right)$	-4μ	-2μ	$\dfrac{\eta}{6}$	0
$16+7\lambda$	$9-2\lambda$	$4-2\lambda$	$1-2\lambda$	$-\lambda$	$-\left(\dfrac{\eta}{6}+8\mu\right)$	-6μ	-4μ	-2μ	$\dfrac{\eta}{6}$
$1+\lambda-\dfrac{\eta}{2}$	$-\left(\lambda-\dfrac{\eta}{2}\right)$	0	0	0	$-\left(\dfrac{\eta}{2}+3\mu\right)$	0	0	0	0
$8+7\lambda-\dfrac{\eta}{2}$	$1-6\lambda$	$-\left(\lambda-\dfrac{\eta}{2}\right)$	0	0	$-(\eta+12\mu)$	-3μ	0	0	0
$27+19\lambda-\dfrac{\eta}{2}$	$8-12\lambda$	$1-6\lambda$	$-\left(\lambda-\dfrac{\eta}{2}\right)$	0	$-\left(\dfrac{3\eta}{2}+27\mu\right)$	-12μ	-3μ	0	0
$64+37\lambda-\dfrac{\eta}{2}$	$27-18\lambda$	$8-12\lambda$	$1-6\lambda$	$-\left(\lambda-\dfrac{\eta}{2}\right)$	$-(2\eta+48\mu)$	-27μ	-12μ	-3μ	0

Figure 4.35 The [A] matrix for a five-girder bridge.

TABLE 4.2 Values of Characterizing Parameters for Various Harmonic Numbers

Parameter	Value of parameter for harmonic no.			
	1	2	3	4
λ	0.061	0.015	0.007	0.004
η	7.766	0.485	0.096	0.030
μ	0.154	0.039	0.017	0.010

It is again recalled that the second equation of the bank of equations defined by Eq. (4.37) may, at choice, be obtained by taking moments about the outer left-hand girder, as was done for developing Eq. (4.27), rather about the outer right-hand girder, as was done for the second equation of Eq. (4.35). If the former procedure is followed, then the second equation of Eq. (4.35) is replaced by

$$\frac{1}{N-1}[-\lambda\rho_1 + \rho_2 + 2\rho_3 + \cdots + (N-2)\rho_{N-1} + \lambda\rho_N$$

$$+ \mu\rho_1^* + \cdots + \mu\rho_N^*] = \frac{e}{N-1} \quad (4.77b)$$

and for the first harmonic the resulting matrix is then as shown in Fig. 4.36. It may be noted that in the computer programs SECAN1 and SECAN2, which are described in Chap. 9, the second equation of the matrix is of this form.

Calculation of vector $\{R\}$. The vector $\{R\}$ is defined by Eqs. (4.37) and (4.38); the instructions for developing its terms are given in App. III. As discussed in Sec. 4.3, this vector can be calculated separately for each longitudinal line of wheels, and the set of equations solved also

1.000	1.000	1.000	1.000	1.000	0.0	0.0	0.0	0.0	0.0
-0.015	0.250	0.500	0.750	1.000	0.039	0.039	0.039	0.039	0.039
1.061	-0.061	0.0	0.0	0.0	-1.603	1.294	0.0	0.0	0.0
4.182	0.879	-0.061	0.0	0.0	-1.911	-0.308	1.294	0.0	0.0
9.303	3.879	0.879	-0.061	0.0	-2.220	-0.617	-0.308	1.294	0.0
16.425	8.879	3.879	0.879	-0.061	-2.528	-0.925	-0.617	-0.308	1.294
-2.823	3.823	0.0	0.0	0.0	-4.346	0.0	0.0	0.0	0.0
4.541	0.636	3.823	0.0	0.0	-9.617	-0.463	0.0	0.0	0.0
24.269	7.272	0.636	3.823	0.0	-15.813	-1.851	-0.463	0.0	0.0
62.361	25.908	7.272	0.636	3.823	-22.935	-4.164	-1.851	-0.463	0.0

Figure 4.36 The $[A]$ matrix corresponding to the first harmonic for the five-girder bridge shown in Fig. 4.34.

$$\begin{Bmatrix} 2.000 \\ 0.081 \\ 1.670 \\ 7.024 \\ 16.377 \\ 29.731 \\ 1.844 \\ 13.885 \\ 47.986 \\ 116.149 \end{Bmatrix}$$

Figure 4.37 The $\{R\}$ vector.

separately for each line of wheels. Alternatively, if the lines of wheels are identical, as they are in the case under consideration, the various $\{R\}$ vectors for the different lines of wheels can be added to give a single vector. By using this vector the set of equations need be solved only once for all lines of wheels. The combined $\{R\}$ vector for the two lines of wheels in Fig. 4.34 is calculated to be as shown in Fig. 4.37.

Solution of equations. After the $[A]$ matrices and $\{R\}$ vectors have been formed, the sets of 10 simultaneous equations are solved separately for each harmonic to obtain $\{\rho\}$ vectors. The resulting $\{\rho\}$ vectors for the first five harmonics are shown in Fig. 4.38. It is noted that the top five coefficients in each vector are used to obtain deflections, bending moments, and shears in the five girders, and the remaining five to obtain twisting moments.

Calculation of girder moments. The formation of matrix $[A]$ and vector $\{R\}$ and then the solution of equations to obtain vector $\{\rho\}$ are repeated for each harmonic in turn. For the example under consideration, the values of the distribution coefficient ρ_1 for girder 1, identified in Fig. 4.34, are found to be 1.278, 1.498, 1.517, 1.498, and 1.473 for the first five harmonics respectively. For a very large harmonic number, say,

$$\begin{Bmatrix} 1.27812 \\ 0.62887 \\ 0.17291 \\ -0.01195 \\ -0.06795 \\ -0.70123 \\ -0.59581 \\ -0.29927 \\ -0.09393 \\ -0.03648 \end{Bmatrix} \quad \begin{Bmatrix} 1.49756 \\ 0.53362 \\ -0.01249 \\ -0.01952 \\ 0.00083 \\ -1.57806 \\ -1.12833 \\ -0.12649 \\ 0.03464 \\ 0.01066 \end{Bmatrix} \quad \begin{Bmatrix} 1.51705 \\ 0.53243 \\ -0.05197 \\ 0.00241 \\ 0.00007 \\ -3.69678 \\ -2.45400 \\ 0.17094 \\ -0.00341 \\ -0.00174 \end{Bmatrix} \quad \begin{Bmatrix} 1.49810 \\ 0.55506 \\ -0.05987 \\ 0.00749 \\ -0.00078 \\ -8.08405 \\ -5.56611 \\ 0.67846 \\ -0.09070 \\ 0.01471 \end{Bmatrix} \quad \begin{Bmatrix} 1.47317 \\ 0.57807 \\ -0.05776 \\ 0.00727 \\ -0.00075 \\ -15.12858 \\ -11.14467 \\ 1.37521 \\ -0.18244 \\ 0.02745 \end{Bmatrix}$$

Harmonic No. 1   Harmonic No. 2   Harmonic No. 3   Harmonic No. 4   Harmonic No. 5

Figure 4.38 The $\{\rho\}$ vectors corresponding to the first five harmonics.

15, which for practical purposes is large enough to correspond to ∞, the $[A]$ is formed and the simultaneous equations are solved to give the following values of distribution coefficients:

$$(\rho_1)_\infty = 1.333$$

$$(\rho_2)_\infty = 0.672$$

$$(\rho_3)_\infty = -0.005 \qquad (4.78)$$

$$(\rho_4)_\infty = 0.000$$

$$(\rho_5)_\infty = 0.000$$

By using the above values the equation for moment in girder 1 can be obtained from Eq. (4.44):

$$M_1 = 1.333 \times M_L - 7.100 \times 10^6(1.333 - 1.278) \sin \frac{\pi x}{L}$$

$$-0.437 \times 10^6(1.333 - 1.498) \sin \frac{2\pi x}{L}$$

$$+ 0.541 \times 10^6(1.333 - 1.517) \sin \frac{3\pi x}{L}$$

$$+ 0.154 \times 10^6(1.333 - 1.498) \sin \frac{4\pi x}{L}$$

$$- 0.069 \times 10^6(1.333 - 1.473) \sin \frac{5\pi x}{L} \qquad (4.79)$$

The moment at girder 1 at midspan is now examined. From simple beam statics the free moment M_L at midspan due to one line of wheels is found to be equal to 7.874×10^6 units as shown in Fig. 4.39. Thus at x equal to 508, M_1 is found to be equal to

$$M_1 = 10.496 \times 10^6 - (0.369 + 0.000 - 0.100$$

$$- 0.000 - 0.010) \times 10^6$$

$$= 10.017 \times 10^6 \qquad (4.80)$$

The second and fourth terms within parentheses are each equal to zero. Other successive terms become smaller and smaller for higher harmonics. It is interesting to note that if only the first harmonic had been used, M_1 would have been equal to 10.127×10^6, which is within about 1 percent of the moment obtained after five harmonics.

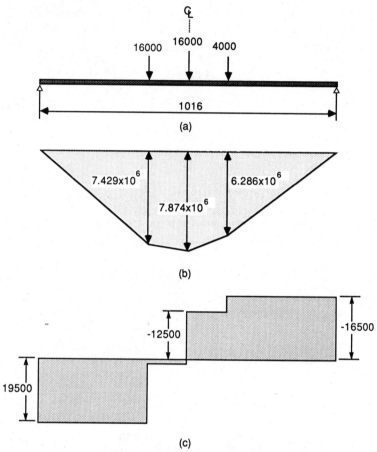

Figure 4.39 Free-moment and shear diagrams due to one line of wheels. (*a*) Beam under three wheels. (*b*) Bending-moment diagram. (*c*) Shear-force diagram.

Midspan moments in the other girders are calculated by considering only the first harmonic.

$$M_2 = 0.672 \times 7.874 \times 10^6 - 7.100 \times 10^6$$
$$\times (0.672 - 0.628) \times 1.0$$
$$= 4.979 \times 10^6 \tag{4.81}$$

Similarly,

$$M_3 = 1.189 \times 10^6$$
$$M_4 = -0.088 \times 10^6 \tag{4.82}$$
$$M_5 = -0.470 \times 10^6$$

Average midspan moment per girder due to the two lines of loads is equal to $2 \times 7.874 \times 10^6/5$, or 3.150×10^6. By dividing the midspan moments M_1 to M_5, calculated above, by this average moment, one obtains the distribution factors for midspan moments. These factors, denoted by $F_1, F_2, \ldots, F_5$ for the five girders, are found to have the following values:

$$F_1 = 3.18$$

$$F_2 = 1.58$$

$$F_3 = 0.38 \tag{4.83}$$

$$F_4 = -0.03$$

$$F_5 = -0.15$$

The values of the distribution factors are not usually required to be calculated. They are calculated above only for later comparison with those for longitudinal shears to demonstrate that longitudinal moments and shears are distributed differently.

Calculation of girder shears. Girder shears, i.e., longitudinal shears, are now calculated at the transverse section for which x is equal to 380. As can be seen in Fig. 4.34, this section is close to an axle. Longitudinal shears can be obtained directly from Eq. (4.46) by inserting the already calculated values of the distribution coefficients and K_1, K_2, etc. The shear in girder 1 is now calculated by considering the first five harmonics. As shown in Fig. 4.39, the free shear V_L at x equal to 380 is 19,500. The quantity (x/L) is equal to 380/1016, or 0.374. From Eq. (4.46),

$$V_1 = 1.333 \times 19,500 - \frac{\pi}{1016} [7.100(1.333 - 1.278) \cos (0.374\pi)$$

$$+ 2 \times 0.437(1.333 - 1.498) \cos (2 \times 0.374\pi)$$

$$-3 \times 0.541(1.333 - 1.517) \cos (3 \times 0.374\pi)$$

$$- 4 \times 0.154(1.333 - 1.498) \cos (4 \times 0.374\pi)$$

$$+ 5 \times 0.069(1.333 - 1.473) \cos (5 \times 0.374\pi)] \times 10^6$$

$$= 25,994 - (466 + 313 - 856 - 3 - 137)$$

$$= 26,211 \tag{4.84}$$

If only the first harmonic had been used, the value of V_1 would have been 25,528, which is within 3 percent of the shear obtained after five harmonics.

The values of shears in the other girders are similarly calculated to give:

$$V_2 = 12,250$$

$$V_3 = 1,146$$

$$V_4 = -51 \tag{4.85}$$

$$V_5 = -561$$

As in the case of longitudinal moments, the distribution factors for longitudinal shears are obtained by dividing the values of shears calculated above by the average shear per girder, which is equal to 7800 units. The values of the distribution factors are found to be

$$F_1 = 3.36$$

$$F_2 = 1.57$$

$$F_3 = 0.15 \tag{4.86}$$

$$F_4 = -0.01$$

$$F_5 = -0.07$$

It can be seen that these distribution factors are not the same as those for longitudinal moments, thus demonstrating that moments and shears are distributed differently and that the semicontinuum method can assess these different distributions.

Calculation of global transverse bending moments. The maximum global transverse moments under a single vehicle are usually caused when the vehicle is centrally placed on the transverse cross section of the bridge. Accordingly, the transverse position of the vehicle is taken to be as shown in Fig. 4.40. The longitudinal position of the vehicle remains the same as shown in Fig. 4.34.

To obtain global transverse moments, loads between girders are transferred as equivalent loads to adjacent girders and the resulting load case is as shown in Fig. 4.41. Vectors (ρ) are calculated separately for loads on girders 2 and 3, each for the first five harmonics. Thus, calculated vectors are shown in Fig. 4.42a and b for loads at girders 2 and 3 respectively. By applying the relevant distribution coefficients to the various harmonics of the loading represented by one line of wheels it is a simple matter to obtain the transverse bending-moment diagram for a unit length of the transverse medium.

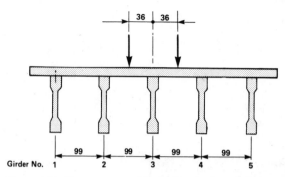

Figure 4.40 Transverse position of loads for maximum transverse moments.

Calculation of transverse shears. The calculation of transverse shears, harmonic by harmonic, follows readily and immediately from the distribution coefficients shown in Fig. 4.38 and the harmonics of the applied external load.

The three concentrated loads due to one line of wheels, shown in Fig. 4.39a, when analyzed into harmonics by using Eq. (2.1), give the following series:

$$
p_x = \frac{32000}{1016} \left(\sin \frac{\pi \times 381}{1016} \sin \frac{\pi x}{L} + \sin \frac{2\pi \times 381}{1016} \sin \frac{2\pi x}{L} + \cdots \right)
$$

$$
+ \frac{32000}{1016} \left(\sin \frac{\pi \times 508}{1016} \sin \frac{\pi x}{L} + \sin \frac{2\pi \times 508}{1016} \sin \frac{2\pi x}{L} + \cdots \right)
$$

$$
+ \frac{8000}{1016} \left(\sin \frac{\pi \times 635}{1016} \sin \frac{\pi x}{L} + \sin \frac{2\pi \times 635}{1016} \sin \frac{2\pi x}{L} + \cdots \right)
$$

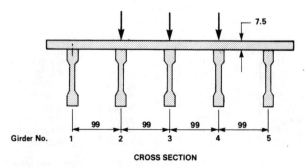

CROSS SECTION

Figure 4.41 Details of apportioned loads to girders for maximum transverse moments.

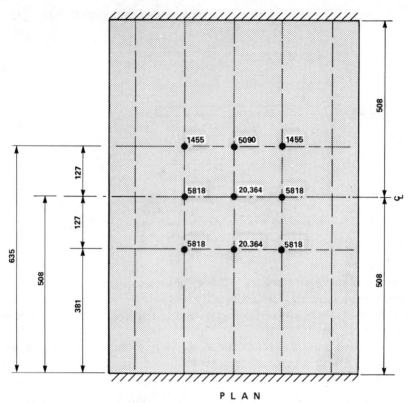

P L A N

Figure 4.41 (*Continued*)

	Harmonic No. 1	Harmonic No. 2	Harmonic No. 3	Harmonic No. 4	Harmonic No. 5
(a)	$\begin{Bmatrix} 0.29028 \\ 0.39677 \\ 0.25191 \\ 0.08383 \\ -0.02280 \\ 0.16046 \\ -0.02058 \\ -0.19081 \\ -0.13226 \\ -0.08854 \end{Bmatrix}$	$\begin{Bmatrix} 0.12637 \\ 0.72333 \\ 0.17553 \\ -0.01977 \\ -0.00547 \\ 0.73168 \\ -0.02119 \\ -0.48938 \\ -0.01745 \\ 0.02433 \end{Bmatrix}$	$\begin{Bmatrix} 0.04871 \\ 0.89490 \\ 0.07039 \\ -0.01525 \\ 0.00126 \\ 0.84127 \\ -0.04611 \\ -0.54248 \\ 0.06712 \\ -0.00569 \end{Bmatrix}$	$\begin{Bmatrix} 0.02139 \\ 0.95504 \\ 0.02881 \\ -0.00591 \\ 0.00068 \\ 0.72610 \\ -0.03558 \\ -0.49415 \\ 0.07111 \\ -0.01403 \end{Bmatrix}$	$\begin{Bmatrix} 0.01085 \\ 0.97762 \\ 0.01359 \\ -0.00232 \\ 0.00026 \\ 0.58690 \\ -0.02199 \\ -0.42631 \\ 0.05615 \\ -0.01006 \end{Bmatrix}$
(b)	$\begin{Bmatrix} 0.05301 \\ 0.25192 \\ 0.39015 \\ 0.25191 \\ 0.05302 \\ 0.18715 \\ 0.20010 \\ -0.00000 \\ -0.20010 \\ -0.18714 \end{Bmatrix}$	$\begin{Bmatrix} -0.02727 \\ 0.17553 \\ 0.70347 \\ 0.17554 \\ -0.02727 \\ 0.05268 \\ 0.47793 \\ -0.00004 \\ -0.47795 \\ -0.05261 \end{Bmatrix}$	$\begin{Bmatrix} -0.01315 \\ 0.07040 \\ 0.88549 \\ 0.07042 \\ -0.01316 \\ -0.09771 \\ 0.55084 \\ -0.00020 \\ -0.55081 \\ 0.09820 \end{Bmatrix}$	$\begin{Bmatrix} -0.00493 \\ 0.02880 \\ 0.95228 \\ 0.02875 \\ -0.00490 \\ -0.10300 \\ 0.50349 \\ 0.00014 \\ -0.50391 \\ 0.10208 \end{Bmatrix}$	$\begin{Bmatrix} -0.00197 \\ 0.01359 \\ 0.97675 \\ 0.01364 \\ -0.00200 \\ -0.07732 \\ 0.43161 \\ -0.00179 \\ -0.42995 \\ 0.07987 \end{Bmatrix}$

Figure 4.42 $\{\rho\}$ vectors corresponding to loading cases in Fig. 4.41. (*a*) Vectors for load on girder 2. (*b*) Vectors for load on girder 3.

That is,

$$p_x = 67.87 \sin \frac{\pi x}{L} + 16.70 \sin \frac{2\pi x}{L} + \cdots \qquad (4.87)$$

For first harmonic responses, we now find the amplitudes of the first harmonic load interactions between the girders and the transverse medium. These amplitudes are shown in Fig. 4.13 as $P_1, P_2, \ldots$, on girders $1, 2, \ldots$, and their values are given by the formulas of Eqs. (4.21), (4.22), and (4.23), so that, for example,

$$P_1 = k_1[(1 + \lambda_1)a_1 - \lambda_1 a_2] \qquad (4.21)$$

In the bridge under consideration all girders are equal and equally spaced so that in Eq. (4.21) λ_1 is replaced by λ. Now by using Eq. (4.33) for $k_1 a_1$ and $k_2 a_2$ and putting $P = 67.87$ from Eq. (4.87), we derive

$$P_1 = 67.87[(1 + \lambda)\rho_1 - \lambda\rho_2] \qquad (4.88)$$

Working to an accuracy of three decimal places and hence inserting $\rho_1 = 1.278$ and $\rho_2 = 0.629$ from the tabulation of first harmonic distribution coefficients in Fig. 4.38 and putting $\lambda = 0.0606$, we find

$$P_1 = 67.87[(1.0606 \times 1.278) - (0.0606 \times 0.629)] = 89.42$$

Similarly,

$$P_2 = 67.87[-\lambda\rho_1 + (1 + 2\lambda)\rho_2 - \lambda\rho_3] = 41.90$$

$$P_3 = 67.87[-\lambda\rho_2 + (1 + 2\lambda)\rho_3 - \lambda\rho_4] = 10.63 \qquad (4.89)$$

$$P_4 = 67.87[-\lambda\rho_3 + (1 + 2\lambda)\rho_4 - \lambda\rho_5] = -1.35$$

$$P_5 = 67.87[-\lambda\rho_4 + (1 + \lambda)\rho_5] = -4.85$$

The load-and-reaction diagram and thence the transverse-shear-force diagram at midspan per unit length are now readily drawn and are shown in Fig. 4.43. It may be noted that the latter diagram, when multiplied by $\sin (\pi x/L)$, gives the transverse shear force per unit length at distance x from the left-hand support.

It may be further noted that the transverse-shear-force diagram shown in Fig. 4.43b may be integrated to give the transverse-bending-moment diagram, including both local and global moments, if desired. Higher harmonics are treated similarly.

A cautionary word is necessary about the use of point loads as representing wheel loads of trains of wheels. This approximation is a very

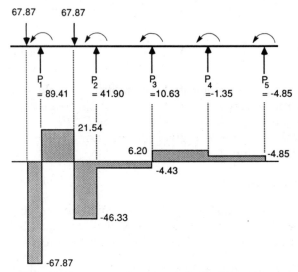

Figure 4.43 Transverse shear due to the first harmonic per unit length at midspan. (*a*) Load-and-reaction diagram. (*b*) Shear-force diagram.

good one for bending effects and is also good for shear-force effects except that it can, of course, lead to very high intensities of shear force in the immediate vicinity of the point load. In order to avoid getting these very high and fictitious local shear intensities it is necessary only to recall the finite size of a tire print and to terminate the harmonic series for shear force at a half wavelength comparable with the length of the print. Alternatively, by using the formulas given in App. I, the point loads can each be replaced by a short patch of uniformly distributed load having the same centroid and the same total load. The effect of this procedure is to make very little change to the lower harmonics of the shear-force series corresponding to the point loads but to take away the intense peaks due to the very high harmonics.

Calculation of longitudinal twisting moments. The calculation of longitudinal twisting moments, harmonic by harmonic, also follows readily and immediately once the distribution coefficients, shown in Fig. 4.38, are available. It is necessary only to substitute the appropriate values into Eqs. (4.66) and (4.67). Thus, in girder 1, for first harmonic effects, the distribution coefficient ρ_1^* is found from Fig. 4.38 to be -0.701, where again an accuracy of three decimal places is used. The amplitude K_1 of the first harmonic of the free-bending-moment diagram is already known to be 7.10×10^6, and all the other quantities needed in the two equations are available. Thence,

$$(f_1)_1 = \frac{(0.154)\pi(99)}{1016} \times -0.701 = -0.033 \tag{4.90}$$

And finally,

$$(T_1)_1 = -0.033 \times 7.10 \times 10^6 \cos \frac{\pi x}{L}$$

$$= -0.235 \times 10^6 \cos \frac{\pi x}{L} \tag{4.91}$$

Other girders and other harmonics are treated similarly, using Eqs. (4.68) and (4.69).

Calculation of transverse twisting moments. To calculate transverse twisting moments, all that is required is to make the appropriate substitutions into Eq. (4.73). Since in this bridge the girders are all equal and equally spaced, Eq. (4.73) takes the simpler form

$$\tau_r = \sum_{m=1,2}^{\infty} (m\pi)K_m\lambda(\rho_{r+1} - \rho_r) \cos \frac{m\pi x}{L} \tag{4.92}$$

In this equation λ is calculated from Eq. (4.39) for each harmonic in turn. For the mth harmonic the value of λ in this bridge is $0.0606/m^2$. Then for the transverse moments between girders 1 and 2 the formula is

$$\tau_1 = \sum_{m=1,2}^{m=\infty} \left(\frac{0.0606}{m}\right) \pi K_m(\rho_{r+1} - \rho_r) \cos \frac{m\pi x}{L} \text{ , giving}$$

$$\tau_1 = 0.0606\pi \left[7.100 \times 10^6(0.629 - 1.278) \cos \frac{\pi x}{L} \right.$$

$$\left. + \frac{0.437 \times 10^6}{2}(0.534 - 1.498) \cos \frac{2\pi x}{L} + \cdots \right]$$

$$= -0.877 \times 10^6 \cos \frac{\pi x}{L}$$

$$- 0.040$$

$$\times 10^6 \cos \frac{2\pi x}{L} + \cdots \tag{4.93}$$

and similarly for the other girders.

Once again it is interesting and instructive to note the dominance of the first harmonic term.

Validity of the method. For the example under consideration, results by the semicontinuum method are compared in Ref. 6 with those obtained by the grillage analogy method. It is noted that the grillage analogy method, which, for example, is dealt with in Ref. 5, is already well tested and validated against other rigorous analytical and experimental methods. Comparisons of distribution factors for midspan girder moments and for shears at three-eighths span are reproduced in Fig. 4.44 for the load case dealt with in this example and two other load cases. The good correspondence between results of the two methods is very encouraging. The very slight difference between the semicontinuum and grillage results, which can be seen in Fig. 4.44, is due to the relatively coarse grillage used in the idealization of the latter method. The differences between the results of the two methods can be expected to decrease with increase in the number of transverse members in the grillage idealization. The semicontinuum idealization incorporates, in effect, an infinite number of transverse members and is therefore preferred from the standpoint of accuracy.

To a purist the representation of distribution factors in a slab-on-girder bridge by a continuous curve may not appear to be correct. Strictly speaking, the factors should have been shown as a series of

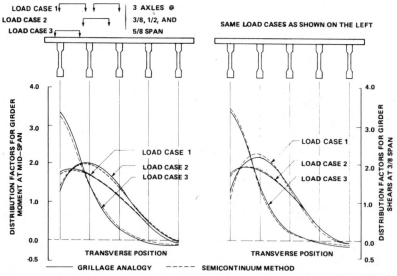

Figure 4.44 Comparison of results by grillage analogy and semicontinuum methods.

points, one under each girder. However, the format used in Fig. 4.44, which is valid for orthotropic plates, was adopted in favor of an easy-to-follow representation.

4.11 Girders with Varying Moments of Inertia

The general semicontinuum method of analysis as developed in this chapter is based upon the assumption of longitudinal girders that are of constant flexural and torsional stiffnesses, EI and GJ, throughout their length. This constancy is the basis for defining the governing nondimensional parameters η, λ, and μ whose values determine the live-load distribution properties of the bridge.

It is fortunate that the general semicontinuum method can also be applied to a bridge whose girders have moments of inertia that vary along the length of the bridge, provided that the variation is moderate. Guidance to the designer on what constitutes moderate variation is given in Chap. 5, and expressions for calculating equivalent constant moments of inertia, for purposes of load distribution analysis, are given in App. IV. The relevant expressions for continuously varying and stepwise-varying cases are given by Eqs. (IV.10) and (IV.15) respectively.

REFERENCES

1. Bakht, B., and Jaeger, L. G.: *Bridge Analysis Simplified,* McGraw-Hill, New York, 1985.
2. Cusens, A. R., and Pama, R. P.: *Bridge Deck Analysis,* Wiley, London, 1975.
3. Hendry, A. W., and Jaeger, L. G.: *The Analysis of Grid Frameworks and Related Structures,* Prentice-Hall, Englewood Cliffs, N.J., 1958.
4. Hewitt, B. E., and Batchelor, B. de V.: Punching shear strength of restrained slabs, *ASCE Journal of the Structural Division,* 101(9), 1975.
5. Jaeger, L. G., and Bakht, B.: The grillage analogy in bridge analysis, *Canadian Journal of Civil Engineering,* 9(2), 1982, pp. 224–235.
6. Jaeger, L. G., and Bakht, B.: Bridge analysis by the semicontinuum method, *Canadian Journal of Civil Engineering,* 12(3), 1985, pp. 573–582.
7. Jaeger, L. G., and Bakht, B.: The use of harmonics in the semi-continuum analysis of bridges, *Proceedings, Canadian Society for Civil Engineering Annual Conference, Saskatoon,* 1985, vol. IIB, pp. 83–98.
8. Rowe, R. E.: *Concrete Bridge Design,* Applied Science Publishers, London, 1962, pp. 62–114.
9. Thompson, W., and Tait, P. C.: *Treatise on Natural Philosophy,* vol. I, part II, Cambridge University Press, London, 1883.

Bridges with Random Intermediate Supports

5.1 Introduction

This chapter deals with the analysis of slab-on-girder bridges with right simply supported ends and arbitrarily placed discrete intermediate supports. An example of the general case is shown in Fig. 5.1a. A continuous-span bridge is a particular case of the general category. An example of this particular case is shown in Fig. 5.1b.

Bridges with intermediate supports are complex by nature and cannot be directly analyzed by the semicontinuum method of Chap. 4. The problem can, however, be considerably simplified by using the force method, which has already been discussed in Chap. 2 in conjunction with the analysis of continuous beams. For bridges with intermediate supports this method is discussed in the following section.

5.2 Force Method

The force method is based on an old technique for dealing with statically indeterminate reactions of intermediate supports; it requires the following steps of calculations:

1. Remove all intermediate supports and, by treating the bridge as simply supported at the two ends, find deflections at the intermediate-

support locations due to the applied loading, using the semicontinuum method of Chap. 4.

2. Again treating the bridge as simply supported at its two ends, find the (usually upward) forces at each of the intermediate-support locations which would bring the bridge at these locations back to their respective original positions.

3. The bridge with intermediate supports can then be analyzed by the semicontinuum method of Chap. 4 as a simply supported bridge which is subjected to downward-applied loads and the usually upward reactions of the intermediate supports calculated above.

It is noted that the force method has been extensively used for solving the problem of bridges with intermediate supports. For example, it was used in Ref. 4 to analyze continuous bridges with negligible torsional stiffness by the semicontinuum method. As noted elsewhere in the book, the semicontinuum method of Chap. 4 can take account of finite torsional stiffnesses of the bridge in both the longitudinal and the transverse directions. The force method has also been successfully used in conjunction with the finite strip method and the orthotropic plate method (Refs. 2 and 3 respectively).

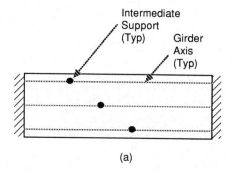

(a)

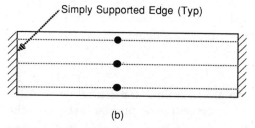

(b)

Figure 5.1 Plans of slab-on-girder bridges with intermediate supports. (*a*) General case with randomly spaced intermediate supports. (*b*) Particular case of a continuous-span bridge.

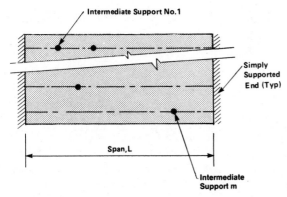

Intermediate Support No.1

Simply
Supported
End (Typ)

Span, L

Intermediate
Support m

Figure 5.2 Plan of a slab-on-girder bridge with m intermediate supports.

Analytical procedure. For developing the analytical method, a general case is considered in which each intermediate support can have a spring support and a prescribed deflection. Of course, the particular cases of unyielding supports and zero prescribed settlements are implicitly covered by the general case. It is assumed that all intermediate supports lie on the girder centerlines. The general case of a bridge with m intermediate supports, as shown in Fig. 5.2, is considered. The deflection flexibilities of the intermediate supports are denoted by $f_1, f_2, \ldots,$ f_m, their prescribed deflections as $\delta_1, \delta_2, \ldots, \delta_m$, and their reactions by $R_1, R_3, \ldots, R_m$. The subscripts to f, δ, and R refer to the support numbers. The notation for support i is illustrated in Fig. 5.3. The deflection of support i due to applied loading is denoted by W_i. It can

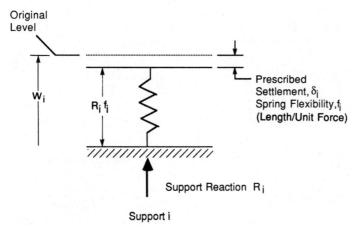

Original
Level

W_i

R_i f_i

Prescribed
Settlement, δ_i
Spring Flexibility, f_i
(Length/Unit Force)

Support Reaction R_i

Support i

Figure 5.3 Notation for support i.

be readily appreciated that this deflection is given by

$$W_i = R_i f_i + \delta_i \tag{5.1}$$

The bridge without the intermediate supports, with span L which corresponds to the distance between the simply supported ends of the bridge, as identified in Fig. 5.2, is now considered. The deflection at the location of intermediate support i due to a unit upward force at the location of intermediate support j is denoted by w_{ij}. Further, the deflection at the location of support i, due to applied loading in the bridge without the intermediate supports, is denoted by $\overline{w}_i$. It can be readily shown that at intermediate support location i

$$W_i = \overline{w}_i - (R_1 w_{i1} + R_2 w_{i2} + \cdots + R_m w_{im}) \tag{5.2}$$

By equating the right-hand sides of Eqs. (5.1) and (5.2) for support 1, there results

$$\overline{w}_1 - \delta_1 = R_1(w_{11} + f_1) + R_2 w_{12} + \cdots + R_m w_{1m} \tag{5.3}$$

A set of m simultaneous equations can now be written for the m intermediate supports to solve for the m unknown reactions R_1, R_2, $\ldots$, R_m. These equations can be written in the following matrix form.

$$\begin{bmatrix} w_{11} + f_1 & w_{12} & \cdot & \cdot & w_{1m} \\ w_{21} & w_{22} + f_2 & \cdot & \cdot & w_{2m} \\ \vdots & \vdots & \vdots & \vdots & \vdots \\ w_{m1} & w_{m2} & \cdot & \cdot & w_{mm} + f_m \end{bmatrix} \begin{Bmatrix} R_1 \\ R_2 \\ \vdots \\ R_m \end{Bmatrix} = \begin{Bmatrix} \overline{w}_1 - \delta_1 \\ \overline{w}_2 - \delta_2 \\ \vdots \\ \overline{w}_m - \delta_m \end{Bmatrix} \tag{5.4}$$

or

$$[A]\{R\} = \{D\} \tag{5.5}$$

The semicontinuum method of Chap. 4 can be applied to the bridge without the intermediate supports to determine the values of $\overline{w}_i$ and w_{ij}. Values of the former are obtained under applied loading, and those of the latter by applying a unit load in turn at the location of each intermediate-support location. Having determined the values of $\overline{w}_i$ and w_{ij} for all intermediate-support locations, the $[A]$ matrix and $\{D\}$ vector can be constructed by also using the prescribed values of f_i and δ_i. The $\{R\}$ vector can readily be obtained by solving Eq. (5.5).

After determining the $\{R\}$ vector, i.e., the unknown intermediate-support reactions, the semicontinuum method of Chap. 4 can again be used for the bridge without intermediate supports to find the effect of the upward-support reactions. The results thus obtained are then superimposed on the corresponding results of the analysis already performed to obtain $\overline{w}_1$, $\overline{w}_2$, $\ldots$, $\overline{w}_m$. The results of the superposition give the solution for the bridge with intermediate supports. A computer program, called SECAN2, has been written to incorporate the above analytical procedure. The description of the FORTRAN program, together with instructions for its use, is given in Chap. 9, and its listings in App. VII. The validity of the program has been established in Ref. 1 by comparing the results of the already verified grillage analogy method (e.g., Ref. 5) with those given by SECAN2.

5.3 Continuous Bridges

As mentioned earlier, a continuous bridge is a particular case of the general category of bridges with randomly spaced intermediate supports. However, continuous bridges are being discussed in this separate section because certain features are peculiar to this form of construction.

Girders with varying moment of inertia. The semicontinuum method given in Sec. 5.2 is applicable only to those bridges in which the longitudinal torsional and flexural ridigities are substantially constant along the length of the bridge. As shown in Fig. 5.4, girder depths of typical continuous-span bridges do vary along the span, as also do the flexural rigidities of the girders. The variation of the flexural rigidity of girders complicates the problem of analysis in two ways, namely,

1. By affecting the total longitudinal moment at a cross section
2. By affecting the manner of transverse distribution of longitudinal responses

These two factors, which fortunately can be handled separately, are discussed below.

Total longitudinal moment across a section. For a given loading on a right simply supported bridge, the total longitudinal moment across a cross section is the same as the corresponding moment in an equivalent

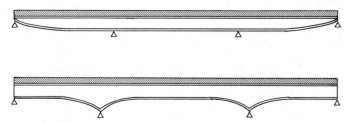

Figure 5.4 Elevations of two continuous-span bridges with variable-depth girders.

beam. This axiomatic phenomenon is illustrated in Fig. 5.5 for a bridge carrying loads on two transverse lines.

As shown with the help of the example of Fig. 5.6, the moments of a two-span continuous beam are, of course, affected by variations of the moment of inertia. For the case shown in the figure, the "haunching" of the beam over the intermediate-support area leads to a decrease of the positive moments and an increase of the negative. The calculation of moments in a continuous beam with varying moment of inertia does not pose a problem; it can be done by simple beam theory. However, whether the bending moment in the beam at a particular location is equal to the sum of the longitudinal moments in the girders of the bridge at that location is another question. The answer to this question is not settled axiomatically by simple statics, as it is in the case of a simply supported bridge. Nevertheless, it is generally assumed in design that the total longitudinal moment across a transverse section due to applied loading on a right continuous bridge is the same as the corresponding beam moment.

This belief is examined with the help of the example of a two-span continuous-slab bridge. As shown in Fig. 5.7a, a concentrated load is applied to one span of this bridge near the intermediate-support line. A fictitious, and undoubtedly exaggerated, pattern of *load dispersion* is shown on the plan of Fig. 5.7a. It follows that the portion of the bridge which lies outside the load dispersion lines does not participate in transferring the applied loading to the slab supports. The portion

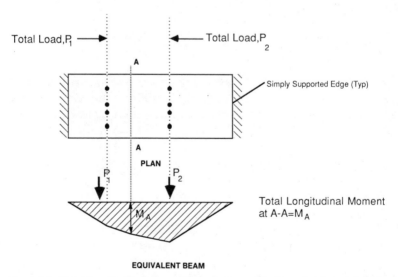

Figure 5.5 Total longitudinal moment at a cross section in a simply supported bridge.

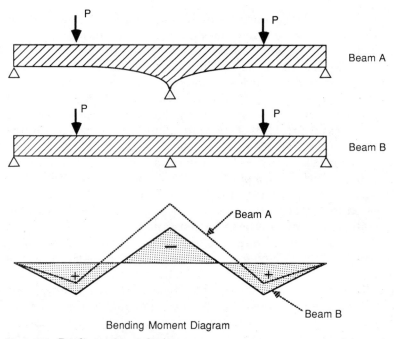

P

P

Beam A

P

P

Beam B

Beam A

−

+

+

Beam B

Bending Moment Diagram

Figure 5.6 Bending-moment diagram.

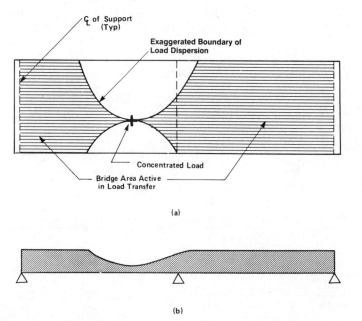

℄ of Support
(Typ)

Exaggerated Boundary of
Load Dispersion

Concentrated Load

Bridge Area Active
in Load Transfer

(a)

(b)

Figure 5.7 Exaggerated representation of total effective flexural rigidity
in a two-span continuous-slab bridge. (a) Plan. (b) Total effective longitudinal flexural rigidity.

of the bridge which does participate in the load transfer is shown shaded in Fig. 5.7a. If the whole bridge does not participate in load transfer, then it is clear that a beam which is equivalent to the bridge will not have a uniform flexural rigidity. The "effective" flexural rigidity for the beam representing the bridge under consideration is shown schematically in Fig. 5.7b. Notwithstanding the exaggerated pattern of load dispersion assumed in Fig. 5.7, it can be readily appreciated that the effective flexural rigidity of the equivalent beam is no longer constant and may vary from load case to load case and that therefore the actual beam moments may deviate from the corresponding moments in the beam of uniform moment of inertia, or flexural rigidity.

It is thus established that the total longitudinal moments at a transverse section in a continuous bridge are not necessarily equal to the corresponding moment in a beam of constant flexural rigidity which represents the full flexural rigidity of the bridge. To study the problem quantitatively, three slab-on-girder bridges, each two-span continuous and having four girders, were analyzed by the grillage analogy method (see, e.g., Ref. 5), and the sum of girder responses thus obtained was compared with the responses given by the beam analysis. The dimensions of the structures were taken to be representative of real-life structures. Details of the load cases considered in the analysis and girder elevations and moments of inertia are shown in Fig. 5.8. As shown in this figure, one of the bridges had girders of uniform moment

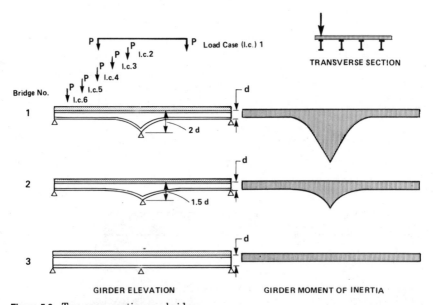

Figure 5.8 Two-span continuous bridges.

of inertia, and the other two had the girder moment of inertia increasing parabolically in different proportions over the intermediate support.

To quantify the deviation from the corresponding beam responses of the total longitudinal responses in the bridge, namely, moments and shears, at a transverse section, a factor δ is defined as

$$\delta = \frac{R_{\text{beam}}}{R_{\text{bridge}}} \tag{5.6}$$

where R_{bridge} is the sum of the girder responses at a transverse section and R_{beam} is the corresponding beam response. Clearly, when δ is equal to 1.0, the sum of the girder responses at a transverse section can be realistically obtained from beam analysis.

The following conclusions were derived from the above analyses:

1. The value of δ was *very* nearly equal to 1.0 for moments and shears in bridge 3 of Fig. 5.8, which has girders of uniform moment of inertia.

2. The value of δ was nearly equal to 1.0 for moments and shears for bridge 2 of Fig. 5.8, in which girder depth over the intermediate support is 1.5 times that of the uniform-depth portion.

3. For bridge 1 of Fig. 5.8, in which the girder depth over the intermediate support is twice that of the uniform-depth portion, the value of δ for both shears and moments varied between 0.89 and 1.15. Specific values of δ for the various load cases for bridge 1 are given in Table 5.1.

4. The value of δ is little affected by the transverse position of the load.

Admittedly, the value of δ for a single concentrated load is too stringent a measure for deciding if the beam analogy can be realistically

TABLE 5.1   Values of δ for Two-Span Bridge 1 Shown in Fig. 5.8

Load case no. (see Fig. 5.8)	Reactions at left-hand support	Reactions at middle support	Reactions at right-hand support	Moment under load	Moment over middle support
1	0.89	1.03	0.89	0.89	1.09
2	0.96	1.01	1.05	0.96	1.05
3	0.92	1.04	1.15	0.92	1.15
4	0.96	1.03	1.09	0.96	1.09
5	1.01	0.99	0.97	1.01	0.97
6	1.01	0.96	0.92	1.01	0.93

NOTE: For a given response, δ = beam response/corresponding sum of girder responses.

used for determining the total responses in a bridge. However, in spite of this very stringent criterion, the bridges with uniform and moderately varying moments of inertia respectively pass the test satisfactorily. It can now be stated with confidence that the beam analogy can be realistically used for these bridges, and consequently the semicontinuum method of Sec. 5.2 may also be used for them. It can be readily appreciated that the value δ, if it is other than 1.0, will converge to 1.0 with an increasing number of loads in the transverse direction. Hence, the values of δ for bridge 1, given in Table 5.1, will move closer to 1.0 for vehicle loads when there are at least two different transverse positions of load. For load cases involving only one vehicle, however, these values may still be so different from 1.0 as to make the use of the beam analogy, and hence the semicontinuum method, inadvisable.

It is concluded, therefore, that, for bridges in which the girder moment of inertia varies as much as it does for bridge 1 of Fig. 5.8, the use of the semicontinuum method may not yield very accurate results.

Equivalent uniform moment of inertia. The load deflection response of a girder of varying moment of inertia may be significantly different from that of an equivalent girder with constant moment of inertia. If this is the case, then a realistic equivalent girder with constant moment of inertia may be impossible to find, which would mean that the semicontinuum method could not be applied to continuous bridges with varying moment of inertia. The first task in the analysis of such bridges is, therefore, to establish whether a realistic equivalent girder can be found. Fortunately, as shown in App. IV, it is possible to represent a girder of even significantly varying moment of inertia by one of constant moment of inertia in such a way that the load deflection characteristics of the two are nearly identical for different patterns of applied loading. This is demonstrated in Fig. IV.5 with reference to specific examples. Appendix IV also gives expressions for calculating the equivalent moment of inertia I_e. Equation (IV.10) provides the expression for calculating I_e for a girder with continuously varying moment of inertia, and Eq. (IV.15) for a girder with stepped moment of inertia. The notation for the former equation is defined in Fig. IV.1, and that for the latter in Fig. IV.4.

Proposed procedure. Based on the above discussion, the following procedure is proposed for the analysis of continuous bridges in which variations of longitudinal flexural rigidity are moderate. (Bridge 1 of Fig. 5.8 is presented as an example in which the variation is not moderate.)

1. Find the moment of inertia I_e of the equivalent girder of constant moment of inertia by the relevant one of Eqs. (IV.10) and (IV.15).

2. Analyze the continuous bridge with equivalent girders by the semicontinuum method of Sec. 5.2.

3. Analyze the continuous bridge as a beam of varying moment of inertia representing the actual bridge and also as a beam of constant equivalent moment of inertia. Denote the ratio of the responses in the beams of the former and the latter respectively as F. For a given position along the beam, the value of F for a certain response, namely, moment, shear, or deflection, is given by

$$F = \frac{R_v}{R_e} \tag{5.7}$$

where R_v and R_e are the responses in beams of varying and equivalent constant moments of inertia respectively.

4. Multiply the responses obtained by the semicontinuum method in Par. 2 by the appropriate value of F to obtain the final values of responses.

5.4 Worked Example

To illustrate the use of the above procedure, a two-span equivalent continuous bridge with four girders is analyzed by the grillage analogy method under two concentrated loads applied at an outer girder. Only girder moments over the intermediate support are investigated. As shown in Fig. 5.9, the loads are symmetrical with respect to the line of intermediate supports. The units of the various quantities are deliberately omitted in order to maintain generality and can be regarded as any mutually compatible set. Some relevant details of the bridge are given in Fig. 5.9; others are presented as follows:

Spans	30, 30
Modulus of elasticity	25.0×10^6
Shear modulus	10.9×10^6
Torsional inertia of a girder (assumed constant)	2.39×10^{-6}

By using the values of the actual moment of inertia at the 11 equally spaced intermediate points, the value of I_e is found to be 0.0046 unit by Eq. (IV.10). This value is also shown in Fig. 5.9.

For the bridge with girders of equivalent constant moment of inertia, the grillage analysis gives the following longitudinal, i.e., girder, mo-

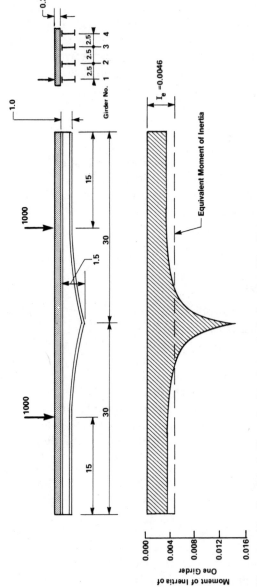

Figure 5.9 Details of the bridge solved in the worked example of Sec. 5.4.

ments over the intermediate supports:

Moment for girder 1	-3561
Moment for girder 2	-2022
Moment for girder 3	-680
Moment for girder 4	637

By beam analysis, the moment over the intermediate support in the beam representing the actual bridge is found to be -7261. Similarly, the moment in the beam representing the equivalent bridge of uniform longitudinal flexural rigidity is found to be -5625 units. Therefore, F for moments over the intermediate support is given by

$$F = \frac{7261}{5625} = 1.291$$

The final values of the moments which are applicable to the actual bridge are obtained by multiplying the moments given by the previous analysis. Thus,

Final moment in girder 1	$-3561 \times 1.291 = -4597$
Final moment in girder 2	$-2022 \times 1.291 = -2610$
Final moment in girder 3	$-680 \times 1.291 = -878$
Final moment in girder 4	$637 \times 1.291 = 822$

Verification of results. The bridge of Fig. 5.9 was also analyzed by the grillage analogy method by idealizing it as an assembly of 4 longitudinal beams of varying flexural rigidity and 11 intermediate transverse beams. The above calculated values of final moments compared extremely well with the latter grillage analysis results. The comparison is shown in Fig. 5.10 in the chart which is labeled $\overline{d}/d = 1.5$.

The other chart in Fig. 5.10 is for the bridge in which the girder depth at the intermediate support is twice that at the abutments. This is the same bridge as that labeled no. 1 in Fig. 5.8, for which the semicontinuum method is of questionable validity. It can be seen that the difference between the true and modified moments is somewhat larger than for the previous bridge but still small enough for the proposed procedure to remain applicable for design purposes.

The analysis of the bridge with uniform longitudinal flexural rigidity by the semicontinuum method gave results which are very close to the corresponding grillage analysis. However, only the grillage analysis results are given in the example to facilitate a direct comparison of the results of analysis in using actual and equivalent girders.

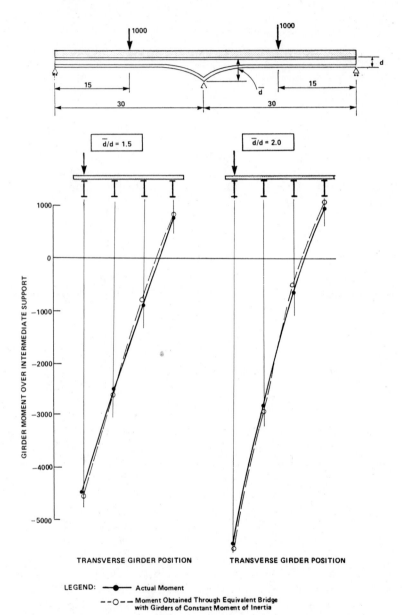

Figure 5.10   Girder moments over intermediate support in two-span continuous bridges.

REFERENCES

1. Bakht, B., and Jaeger, L. G.: Analysis of bridges with intermediate supports by the semi-continuum method, *Proceedings, Annual Conference of the Canadian Society for Civil Engineering, Toronto,* 1986.
2. Cheung, Y. K.: *Finite Strip Method in Structural Analysis,* Pergamon, Oxford, 1976.
3. Cusens, A. R., and Pama, R. P.: *Bridge Deck Analysis,* Wiley, London, 1975.
4. Hendry, A. W., and Jaeger, L. G.: The load distribution in interconnected bridge girders with special reference to continuous beams, *Publications, IABSE,* 15, 1955, pp. 95–115.
5. Jaeger, L. G., and Bakht, B.: The grillage analogy in bridge analysis, *Canadian Journal of Civil Engineering,* 9(2), 1982, pp. 224–235.

Skew Bridges

6.1 Introduction

For bridges in which the planform is a parallelogram, as shown in Fig. 6.1, the angle obtained by subtracting the acute angle of the parallelogram from 90° is termed the *skew angle* of the bridge. Nonskew bridges, i.e., those with zero angle of skew, are also called right bridges. As shown in Fig. 6.1, the span of a skew bridge measured along an unsupported edge of the bridge in plan is called the *skew span,* and the perpendicular distance between the two lines of supports is called the *right span.* The directions parallel with and perpendicular to the flow of traffic on the bridge are still called the longitudinal and transverse directions respectively.

All the semicontinuum methods of analysis presented in this book, except those given in Sec. 6.3, deal only with right bridges. However, as shown in Sec. 6.2, these methods can also be utilized for the analysis of certain skew bridges of the slab-on-girder type of construction.

The semicontinuum method is extended in Sec. 6.3 to cover the specific problem of skew bridges with three girders.

6.2 Approximate Analysis of Skew Slab-on-Girder Bridges

When the skew angle of a bridge is small, say, less than 20°, it is frequently considered safe to ignore the angle of skew and to analyze the bridge as a right bridge whose span is equal to the skew span, as shown in Fig. 6.2. This procedure of analyzing skew bridges is deemed to be safe on the supposition that it leads to conservatively safe estimates of longitudinal moments and shears for the skew bridge. The

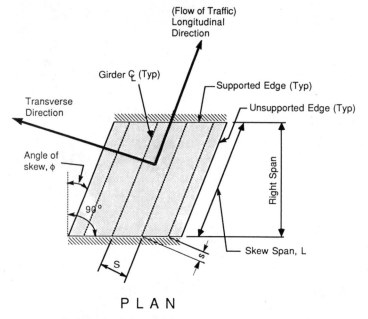

Figure 6.1 Definition of skew-bridge terms.

shortcomings of this procedure have been that, even if it gives safe estimates of the required responses, it is difficult for the engineer to establish the degree of conservatism involved and that twisting moments are not directly considered.

Some codes do not permit the use of the approximate procedure discussed above for bridges with larger angles of skew. The imposition of such a limit on the angle of skew, in relation to certain methods of analysis, implies that the differences between skew-bridge responses and the corresponding responses of the equivalent right bridge depend

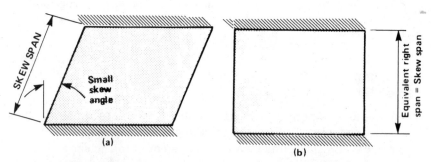

Figure 6.2 Usual method of analyzing skew bridges as right bridges. (*a*) Plan of skew bridge. (*b*) Plan of equivalent right bridge.

solely on the angle of skew. It is shown in this section that girder spacing and bridge span also have an influence on these differences.

6.2.1 Mechanics of load distribution

The mechanics of transverse distribution of loads between the girders of a skew slab-on-girder bridge can be studied with the simple example of a bridge with four torsionally weak girders. As shown in Fig. 6.3*a*, two adjacent girders of this bridge are subjected to four concentrated loads which form a rectangular pattern in plan. If the skew bridge is analyzed as an equivalent right bridge, the four concentrated loads would assume the positions shown in Fig. 6.3*b*.

A study of Fig. 6.3*a* and *b* readily leads to the conclusion that the differences in the girder responses of the skew and equivalent right bridges depend upon two factors. One factor relates to the relative longitudinal positions of loads, and the other to the differences in effective flexural rigidities of girders. Fortunately, the two factors are not related and can be dealt with separately.

Relative longitudinal load positions. The planform of the equivalent right bridge is obtained by conceptually distorting the plan of the skew bridge so that it becomes right. The orthogonal pattern of the four concentrated loads on the plan is maintained on the equivalent right bridge by fixing the relative positions of loads with respect to only one of the girders, namely, the left-hand outer girder. An examination of Fig. 6.3*a* and *b* will readily show that this procedure changes the

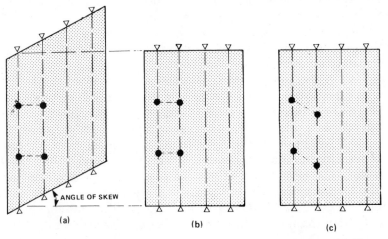

Figure 6.3 Load positions on the plan of right and skew bridges. (*a*) Skew bridge. (*b*) Right bridge with orthogonal pattern of loads. (*c*) Right bridge with skew pattern of loads.

relative position of loads on the second loaded girder. It is obvious that the moments in this second girder will not be realistically represented by the idealization of Fig. 6.3b.

By contrast with the procedure that led to Fig. 6.3b, when the plan of the skew bridge is conceptually deformed to become right, the same deformation is now applied to the load positions. Thus, this revised procedure leads to Fig. 6.3c, in which the pattern of the four concentrated loads, in plan, no longer is orthogonal but is skew. It can be seen that in this conceptual bridge the longitudinal positions of the loads relative to the supports of the girders on which they are applied are the same as in the skew bridge. Now consider the case in which the load distribution characteristics of the bridge are so poor that the applied loads are retained almost entirely by the externally loaded girders. It can be readily seen that in this case the girder responses of the conceptual right bridge of Fig. 6.3c would be the same as the corresponding responses of the actual skew bridge. The same, of course, cannot be said about the conceptual right bridge of Fig. 6.3b. Thus it can be concluded that the preferred equivalent "right bridge" formulation is the one shown in Fig. 6.3c, in which the load positions in plan are conceptually deformed so that they form a skew pattern.

Influence of load distribution characteristics.   When the load distribution characteristics of the bridge under consideration are other than very poor, even the better simplification of Fig. 6.3c may lead to erroneous estimates of the girder responses. This kind of likely error is due to the differences between effective flexural rigidities of the girders of the skew and equivalent right bridges. This aspect of load distribution in skew bridges is studied in the following paragraphs with the help of a simplistic model.

Consider a four-girder skew bridge carrying a concentrated load at the midspan of an outer girder. As shown in Fig. 6.4a, this bridge is idealized as a grid of four longitudinal and one transverse beams. The transverse beam, which passes through the load and is at right angles to the longitudinal beams, represents the deck-slab area shown shaded in this figure. The plan of the grid is now conceptually deformed to become right. As shown in Fig. 6.4b, the transverse beam now intersects all the longitudinal beams at the midspan. Clearly, the transposition does not alter the longitudinal position of the load on the externally loaded girder. Therefore, any differences between the corresponding longitudinal-beam responses of the two idealizations should result only from changes in the load distribution characteristics of the two grids.

The transfer of loads from the externally loaded girder to the other girders is influenced by the relative flexural rigidities of the various

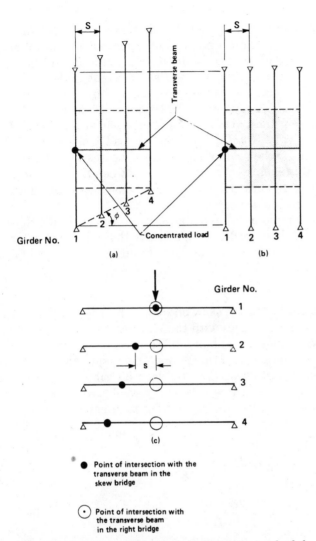

Figure 6.4 Simplistic model to study load retention in a loaded girder. (*a*) Plan of a skew grillage. (*b*) Plan of the equivalent right grillage. (*c*) Girders considered separately.

girders. For the two grids of Fig. 6.4*a* and *b*, the load transfer between the longitudinal beams can take place only through the transverse beam. As shown in Fig. 6.4*c*, the transverse beam of the right grid intersects all the longitudinal beams at their midspans, but the transverse beam of the skew grid intersects only the externally loaded beam at the midspan. The result of relocating the points of load transfer on the longitudinal beams is to change the deflections at load transfer points and consequently to change the effective flexural rigidities of the longitudinal beams. It can be appreciated that, owing to the change in the positions of the load transfer points on the various longitudinal beams, the loads retained by the externally loaded longitudinal beams would not be the same in the skew and equivalent right grids.

The relocation of the points of load transfer, defined by the distance *s* shown in Fig. 6.4*c*, depends not only upon the angle of skew but also on the spacing of the longitudinal beams. Further, it can be appreciated that the effect of the relocation of the point of load transfer would become less significant if the span of longitudinal beams were larger. Another factor which influences the differences between the behaviors of the two grids is the flexural rigidity of the transverse beam with respect to that of longitudinal beams.

From the simple example presented above, it can be concluded that the skew angle of a slab-on-girder bridge is not the only parameter which causes the load distribution properties to differ from those of a right bridge.

Quantification of load distribution in skew bridges.   The semicontinuum method for analyzing skew bridges with three girders, as presented in Sec. 6.3, can be conveniently used to quantify differences in the girder responses of skew slab-on-girder bridges and their equivalent right counterparts. As shown in that section, the load distribution characteristics of skew bridges with three torsionally weak girders depend entirely upon two dimensionless characterizing parameters, η and ϵ, which are defined as follows.

$$\eta = \frac{12}{\pi^4} \left(\frac{L}{S}\right)^3 \frac{LD_y}{EI} \qquad (6.1)$$

$$\epsilon = \frac{2S \tan \phi}{L} \qquad (6.2)$$

where the notation is the same as used in Chap. 4 and as illustrated in Fig. 6.1.

It was shown earlier that the two factors responsible for differences in girder responses of the skew and equivalent right bridges are the quantity s ($= S \tan \phi$), the skew span L, and the flexural rigidity of the deck slab (represented by D_y) relative to the flexural rigidity EI of each girder. The fact that all these factors are implicit in the two characterizing parameters η and ϵ leads to the conclusion that these two parameters may also be applicable to skew bridges with more than three girders.

6.2.2 Estimation of error

The degree of error incurred by analyzing a skew bridge as right can conveniently be considered in terms of a factor δ which is defined by

$$\delta = \frac{(R)_\epsilon - (R)_{\epsilon=0}}{(R)_{\epsilon=0}} \times 100 \tag{6.3}$$

where R is the longitudinal response, i.e., bending moment or shear, and the subscript outside the parentheses refers to the value of ϵ used in the analysis. For a right bridge, $\epsilon = 0$. Thus δ gives the percentage difference between the skew- and equivalent right-bridge responses.

Three-girder skew bridges covering the entire practical range of η and ϵ and their equivalent right counterparts were analyzed by the semicontinuum method of Sec. 6.3 for first harmonic loads on the middle girder and one outer girder. This loading is considered representative of real-life conditions in which the applied loading covers a substantial portion of the bridge width. The longitudinal positions of loads on both girders were kept the same in the skew and right bridges. From the results of the analyses mentioned above, values of δ were calculated for maximum moments and shears in the outer and middle girders. The following observations can be made regarding these values of δ.

1. The values of δ were negative for longitudinal moments in all cases, indicating that the process of analyzing a skew slab-on-girder bridge as right is safe insofar as longitudinal, or girder, moments are concerned.

2. For longitudinal shears, the values of δ were nearly always positive, with their absolute values being larger than the corresponding absolute values for longitudinal moments. This reverse trend of errors shows that, for obtaining longitudinal shears, the practice of analyzing skew slab-on-girder bridges as right is likely to lead to errors on the unsafe side.

3. The absolute values of δ for an outer girder, for both moments and shears, were always smaller than those for the middle girder.

The above observations, as shown later, were also found to be valid for bridges with more than three girders, which were analyzed by the grillage analogy method.

For the development of the procedure given later in the section for assessing δ, it is necessary to develop a relationship between η and L. This is done as follows.

By denoting EI/S as D_x, the longitudinal flexural rigidity per unit width, Eq. (6.1) can be rewritten as

$$\eta = \frac{12}{\pi^4} \left(\frac{L}{S}\right)^4 \frac{D_y}{D_x} \tag{6.4}$$

It has been shown in Ref. 2 that for slab-on-girder bridges in North America the upper-bound values of D_x are related to the span length L by

$$D_x = 2257L^2 + 59,575L \tag{6.5}$$

and the lower-bound values of D_x are given by

$$D_x = 1790L^2 + 9250L \tag{6.6}$$

where L is in meters and D_x in kilonewton meters. By using Eqs. (6.4), (6.5), and (6.6), the upper- and lower-bound values of η can be readily obtained; these two bounds are plotted in Fig. 6.5 along with the median values of η against L.

A value of δ corresponding to a given set of values of η and ϵ can be readily transcribed on the $(L, S \tan \phi)$ space by using Eq. (6.2) and

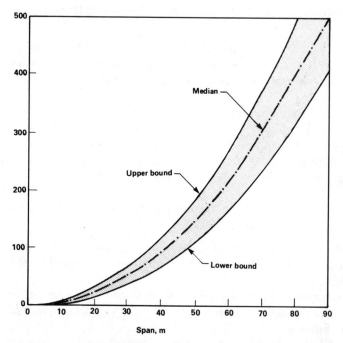

Figure 6.5 Range of η for first harmonic plotted against span length. (η for harmonic m is obtained by dividing the values shown by m^4.)

the median values of η plotted against L in Fig. 6.5. The values of δ for moments in the middle girder, obtained in this way, are plotted in Fig. 6.6 on the $(L, S \tan \phi)$ space covered by practical bridges. The emerging trend indeed confirms engineering judgment, using which it could have been predicted that the influence of the angle of skew diminishes as the span length increases. The value of $S \tan \phi$, which is the same as the offset distance s shown in Figs. 6.1 and 6.4, can be directly computed from the values of S and ϕ or, alternatively, can be read directly from the chart given in Fig. 6.7.

Strictly speaking, the contours of δ given in Fig. 6.6 apply only to bridges with three girders carrying loads on the middle girder and one of the outer girders. In a general way, however, as is verified later in this section, they are also applicable to practical loading cases on bridges with more than three girders.

Figure 6.6 provides a ready means of assessing the degree of error in obtaining maximum longitudinal moments which results from analyzing a skew bridge as right, with the loads being transformed ac-

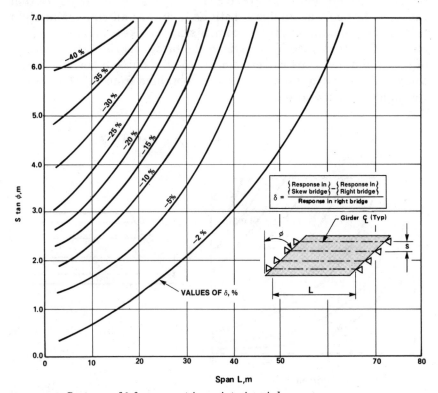

Figure 6.6 Contours of δ for moment in an interior girder.

cording to the scheme of Fig. 6.3c. It should be noted, however, that:

1. A value of δ obtained from this figure applies to an interior girder.
2. The value of δ for an outer girder is also negative but with a smaller absolute value.
3. These values of δ do not apply to longitudinal shear, for the maximum values of which δ is nearly always positive and with a larger absolute value than that for maximum longitudinal moments.

Comparison with grillage analysis. A study of several five-girder skew bridges and their equivalent right counterparts was carried out by grillage analogy. The behavior of these bridges was considered under the three-axle AASHTO [1] MS 20 vehicle. For right bridges, the load positions were established according to the preferred scheme of Fig. 6.3c. This study confirmed that the conclusions arrived at by the semicontinuum analysis of three-girder skew bridges, mentioned earlier, are also valid for bridges with more girders.

As an example, the results of the grillage analysis on a bridge with 45° skew are presented in Fig. 6.8. It can be seen in the figure that the skew bridge was analyzed both as is and as an equivalent right bridge. The maximum girder moments in the skew bridge are quite close to those obtained by analyzing the bridge as right; the two sets of maximum moments correspond to a δ of about −5 percent, which compares well with a δ of −4 percent obtained from Fig. 6.6 for the case under consideration.

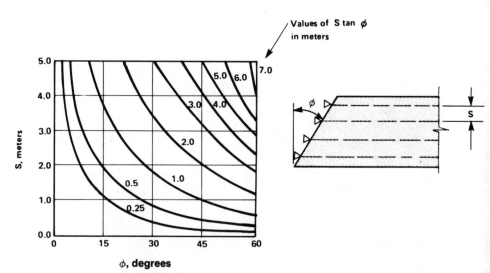

Figure 6.7 Contours of S tan ϕ.

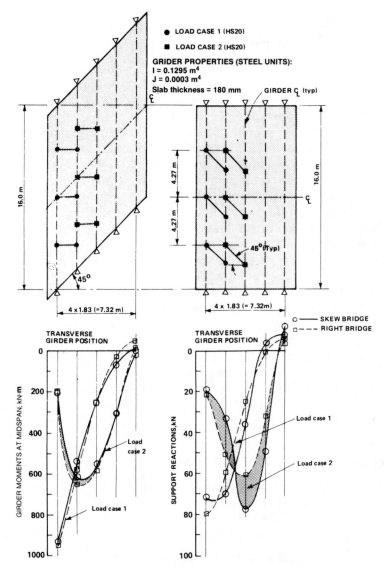

Figure 6.8 Girder moments at midspan and support reactions in a 45° skew bridge.

Figure 6.8 also shows the comparisons of the support reactions in the skew and equivalent right bridges. It can be seen that the maximum support reactions in the skew bridge tend to be larger than the corresponding reactions in the right bridge. These results corroborate the observation made earlier that the process of determining longitudinal shears by treating the skew bridge as right leads to unsafe errors.

6.2.3 Recommendations

The conclusions and recommended practice regarding the analysis of skew bridges as right can be summarized as follows.

1. It is clear from the studies discussed above that the dimensionless ratio $(S \tan \phi)/L$, rather than ϕ alone, should be taken as the appropriate measure of the skewness of a bridge.

2. When a skew bridge is analyzed as an equivalent right bridge, the preferred arrangement of loads on the plan is as shown schematically in Fig. 6.3c. It is noted that this procedure will lead to dissimilar longitudinal lines of loads on the right bridge even if they are similar on the skew bridge. If a semicontinuum analysis program is limited to dealing with only similar longitudinal lines of loads, it can still be used by dealing with each line of loads separately and finally adding all the corresponding results.

3. When the simplification given above is followed, the resulting maximum bending moments are always smaller than the actual moments. The degree of error involved can be approximately determined from the chart of Fig. 6.6 corresponding to the values of span L and $S \tan \phi$ of the bridge.

4. The simplification always leads to unsafe errors in the estimation of maximum longitudinal shears in internal girders. If unsafe errors of up to about 5 percent are considered to be permissible, those skew bridges may be analyzed as equivalent right bridges for which the ratio $(S \tan \phi)/L$ does not exceed about 0.056.

6.3 Analysis of Three-Girder Skew Bridges

The scope of the semicontinuum method of analysis is extended in this section to cover the case of three-girder skew bridges; these are quite common in some parts of the world, e.g., the Indian subcontinent.

6.3.1 Basis of proposed method

Figure 6.9 shows the plan of a skew bridge with three similar, equally spaced girders. As in the case of right bridges, the direction along the girders is referred to as longitudinal and the perpendicular direction as transverse. As shown in Fig. 6.9, the girders, which are spaced a perpendicular distance S apart, are each of length L with coordinates x_1, x_2, and x_3 measured from their left-hand respective supports. The skew angle ϕ of the bridge is defined as

$$\phi = \tan^{-1} \frac{s}{S} \tag{6.7}$$

where s is the offset of a girder support in the longitudinal direction with respect to the corresponding support of an adjacent girder. Hence, as shown in Fig. 6.9, a transverse line CBA starting at $x_1 = L$ intersects girder 2 at $x_2 = L - s$ and girder 3 at $x_3 = L - 2s$. Similarly, the transverse line DEF starting at $x_3 = 0$ intersects girder 2 at $x_2 = s$ and girder 1 at $x_1 = 2s$.

A simplifying assumption is now made which has been justified by rigorous analyses. It is assumed that the load distribution action of the deck slab is limited to the rectangular area identified as $ABCDEF$ in Fig. 6.9. This rectangular area of the deck slab is represented conceptually by an infinite number of transverse beams, constituting the usual transverse medium of the semicontinuum idealization. A portion of this medium contained between x_1 and $(x_1 + \delta x_1)$ has the bending stiffness $D_y\,\delta x_1$. The distribution effects of the deck slab are thus assumed to be limited to values of x_1 from $2s$ to L, x_2 from s to $(L - s)$, and x_3 from 0 to $(L - 2s)$. For the transverse strip δx shown, the position coordinates of the three girders have the following relationships:

$$x_2 = x_1 - s$$
$$x_3 = x_1 - 2s$$

(6.8)

It may be noted that for a skewless (right) bridge s is equal to zero, and the proposed method of analysis then becomes identical to the semicontinuum method for right bridges. Pioneering work on the application of the semicontinuum method to skew bridges is reported in Refs. 5 and 6.

For simplicity, it is assumed that the interaction between the girders and the transverse medium is in the form of vertical forces only. This assumption is equivalent to ignoring the torsional stiffnesses of the

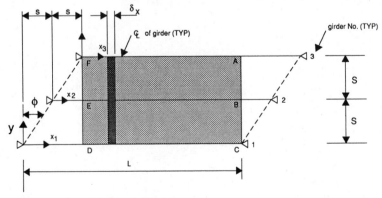

Figure 6.9 Plan of a skew bridge.

girders and the transverse medium and has the effect of slightly over-estimating the maximum responses taken by the girders and the trans-verse medium. Fortunately, the degree of overestimation is not serious.

Representation of loads by harmonics.   By using the procedure that led to Eq. (4.41), free moments M_{Li} on a girder i due to any set of loads can be represented by

$$M_{Li} = M_{Li}^{(1)} \sin \frac{\pi x_i}{L} + M_{Li}^{(2)} \sin \frac{2\pi x_i}{L} + M_{Li}^{(3)} \sin \frac{3\pi x_i}{L} + \cdots \quad (6.9)$$

For the case of a longitudinal beam subjected to N concentrated loads, $M_{Li}^{(1)}$ and $M_{Li}^{(2)}$ are given by

$$
\begin{aligned}
M_{Li}^{(1)} &= \sum_{p=1}^{N} \frac{2 W_p L}{\pi^2} \sin \frac{\pi c_p}{L} \\
M_{Li}^{(2)} &= \sum_{p=1}^{N} \frac{1}{2} \frac{W_p L}{\pi^2} \sin \frac{2\pi c_p}{L}
\end{aligned}
\quad (6.10)
$$

where the notation is as illustrated in Fig. 6.10.

The semicontinuum method of Chap. 4 is general in the sense that it can be applied by considering any number of harmonics. The proposed method for three-girder skew bridges, however, is limited to using only two harmonics. The decision to use only two terms of the harmonic series stems from the following two simplifying assumptions.

1. Third and higher harmonics of bending moments are entirely retained by the externally loaded girder.
2. Third and higher harmonics of deflections are negligibly small.

As has been demonstrated in Chap. 4, these assumptions are quite realistic.

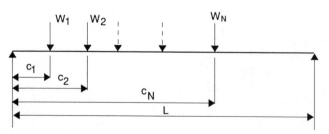

Figure 6.10 N-point loads on a simply supported beam.

6.3.2 Details of analysis

Externally applied loadings on the three girders are defined as P_{L1}, P_{L2}, and P_{L3}, where the second subscript refers to the girder numbers identified in Fig. 6.9. Let the externally applied loading be analyzed into harmonic components as follows:

$$P_{L1} = P_{L1}^{(1)} \sin \frac{\pi x_1}{L} + P_{L1}^{(2)} \sin \frac{2\pi x_1}{L} + \cdots$$

$$P_{L2} = P_{L2}^{(1)} \sin \frac{\pi x_2}{L} + P_{L2}^{(2)} \sin \frac{2\pi x_2}{L} + \cdots \qquad (6.11)$$

$$P_{L3} = P_{L3}^{(1)} \sin \frac{\pi x_3}{L} + P_{L3}^{(2)} \sin \frac{2\pi x_3}{L} + \cdots$$

Let deflections w_1, w_2, and w_3 of the three girders be given by

$$w_1 = a_{11} \sin \frac{\pi x_1}{L} + a_{12} \sin \frac{2\pi x_1}{L}$$

$$w_2 = a_{21} \sin \frac{\pi x_2}{L} + a_{22} \sin \frac{2\pi x_2}{L} \qquad (6.12)$$

$$w_3 = a_{31} \sin \frac{\pi x_3}{L} + a_{32} \sin \frac{2\pi x_3}{L}$$

It is to be noted that, consistently with the simplifying assumptions made above, the deflections are not shown in Eqs. (6.12) as infinite series but as being composed of only two terms.

The upward reactions on the three girders due to their interaction with the deck slab are denoted as R_1, R_2, and R_3 respectively; these reactions can be shown to be represented by

$$R_1 = R_1^{(1)} \sin \frac{\pi x_1}{L} + R_1^{(2)} \sin \frac{2\pi x_1}{L}$$

$$R_2 = R_2^{(1)} \sin \frac{\pi x_2}{L} + R_2^{(2)} \sin \frac{2\pi x_2}{L} \qquad (6.13)$$

$$R_3 = R_3^{(1)} \sin \frac{\pi x_3}{L} + R_3^{(2)} \sin \frac{2\pi x_3}{L}$$

A transverse strip of unit width of the deck slab is now considered in isolation. As shown in Fig. 6.11, the strip is subjected, at the girder locations, to applied loading intensities P_{L1}, P_{L2}, and P_{L3} and to reaction intensities R_1, R_2, and R_3. The net upward loading Q_1 on the transverse strip at the location of girder 1 has first and second harmonic components with amplitudes $Q_1^{(1)}$ and $Q_1^{(2)}$ respectively, which are given by

$$Q_1^{(1)} = R_1^{(1)} - P_{L1}^{(1)}$$
$$Q_1^{(2)} = R_1^{(2)} - P_{L1}^{(2)}$$
(6.14)

Similarly, the net upward force Q_2 at the location of girder 2 has components $Q_2^{(1)}$ and $Q_2^{(2)}$, and at the location of girder 3 Q_3 is composed of $Q_3^{(1)}$ and $Q_3^{(2)}$. The net upward forces of the transverse strip of unit width of the deck slab are shown in Fig. 6.12 together with the deflections of the strip at the girder locations. The net upward forces are given by the following equations:

$$Q_1 = Q_1^{(1)} \sin \frac{\pi x_1}{L} + Q_1^{(2)} \sin \frac{2\pi x_1}{L}$$

$$Q_2 = Q_2^{(1)} \sin \frac{\pi x_2}{L} + Q_2^{(2)} \sin \frac{2\pi x_2}{L}$$
(6.15)

$$Q_3 = Q_3^{(1)} \sin \frac{\pi x_3}{L} + Q_3^{(2)} \sin \frac{2\pi x_3}{L}$$

By measuring a y coordinate from left to right as shown in Fig. 6.12, the moment-curvature relationship for the transverse strip of unit width of the deck slab gives

$$-D_y \frac{d^2w}{dy^2} = Q_1 y + Q_2[y - S]$$
(6.16)

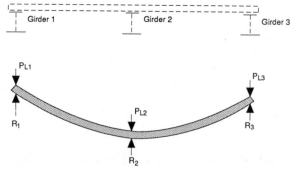

Figure 6.11   Forces on a transverse strip of the deck slab.

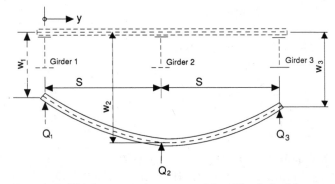

Figure 6.12 Net forces on a transverse strip of the deck slab.

where the bracketed term on the right-hand side is put to zero when it is negative. Integrating Eq. (6.16) twice with respect to y gives

$$D_y w = Q_1 \frac{y^3}{6} + \frac{Q_2}{6} [y - S]^3 + By + C \qquad (6.17)$$

where B and C are constants of integration. The condition $w = w_1$ at $y = 0$ gives

$$C = -D_y w_1 \qquad (6.18)$$

The constant B can be eliminated by putting $w = w_2$ at $y = S$ and $w = w_3$ at $y = 2S$ to give

$$-D_y(w_1 - 2w_2 + w_3) = Q_1 S^3 + Q_2 \frac{S^3}{6} \qquad (6.19)$$

Equation (6.19) can be rearranged to give

$$6Q_1 + Q_2 = \frac{6D_y}{S^3} (-w_1 + 2w_2 - w_3) \qquad (6.20)$$

Vertical equilibrium and moment equilibrium of the transverse strip give respectively

$$Q_1 + Q_2 + Q_3 = 0$$
$$2Q_1 + Q_2 = 0 \qquad (6.21)$$

Equations (6.20) and (6.21) are solved to give

$$Q_1 = \frac{3D_y}{2S^3}(-w_1 + 2w_2 - w_3) \tag{6.22}$$

$$Q_2 = \frac{-3D_y}{S^3}(-w_1 + 2w_2 - w_3) \tag{6.23}$$

$$Q_3 = Q_1 \tag{6.24}$$

The fact that Q_3 is equal to Q_1, even for eccentric loads, may appear contrary to cursory judgment. However, this outcome will become intuitively justifiable if it is noted that Q is, in fact, the *net* upward force. For convenience, the following terms are defined.

$$k = \frac{\pi^4 EI}{L^4} \tag{6.25}$$

$$\eta = \frac{12}{\pi^4}\left(\frac{L}{S}\right)^3 \frac{LD_y}{EI} \tag{6.26}$$

By using Eqs. (6.8), (6.12), and (6.26), Eq. (6.22) can be rewritten as

$$\begin{aligned}
Q_1 = \frac{\eta k}{8}\bigg\{ &-a_{11}\sin\frac{\pi x_1}{L} - a_{12}\sin\frac{2\pi x_1}{L} \\
&+ 2\left[a_{21}\sin\frac{\pi(x_1 - s)}{L} + a_{22}\sin\frac{2\pi(x_1 - s)}{L}\right] \\
&- a_{31}\sin\frac{\pi(x_1 - 2s)}{L} - a_{32}\sin\frac{2\pi(x_1 - 2s)}{L}\bigg\}
\end{aligned} \tag{6.27}$$

First harmonic load and response of girder 1. The total load on girder 1 comprises the external load P_{L1} and the reactive force Q_1. The former acts over the entire length of the girder, i.e., from $x_1 = 0$ to L, as shown in Fig. 6.13b; its amplitude is a known quantity given by the first equation of Eq. (6.11). The reactive force Q_1, which is given by Eq. (6.27), is valid over portion $2s \leq x_1 \leq L$ as shown in Fig. 6.13c. Also as shown in this figure, the reactive force Q_1, covering only a partial length of the girder, has first and second harmonic components which are continuous over the entire length of the girder.

As shown in App. I, the amplitude of the first harmonic representation of reactive force Q_1 is $2/L \int_{2s}^{L} Q_1 \sin(\pi x_1/L)\, dx_1$.

The first harmonic of the deflection of girder 1 is $a_{11} \sin (\pi x_1)/L$, as may be seen from Eq. (6.12). To produce this deflection, the girder has to be subjected to a first harmonic load $k_1 a_{11} \sin (\pi x_1)/L$. Equating this load to the first harmonic of the total load on the girder gives

$$k\, a_{11} = P_{L1}^{(1)} + \frac{2}{L} \int_{2s}^{L} Q_1 \sin \frac{\pi x_1}{L} \, dx_1 \qquad (6.28)$$

The term *free deflection* is used here for the deflection of a girder if it were to sustain all its applied loads alone without passing any portion of them onto the other girders. The amplitude of the first harmonic component of the free deflection of girder 1 is denoted as A_{11}, so that

$$P_L^{(1)} = k\, A_{11} \qquad (6.29)$$

On substituting for $P_{L1}^{(1)}$ and Q_1 from Eqs. (6.29) and (6.27) respec-

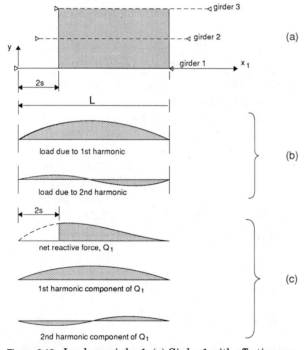

Figure 6.13 Loads on girder 1. (*a*) Girder 1 with effective portion of the transverse medium. (*b*) Two harmonic components of applied load on girder 1. (*c*) Net reactive force Q_1 and its two harmonic components.

tively,

$$a_{11} = A_{11} + \frac{\eta}{8} \left\{ -a_{11} \left(1 - \epsilon + \frac{1}{\pi} \sin 2\pi\epsilon \right) \right.$$

$$- a_{12} \left(\frac{1}{3\pi} \sin 3\pi\epsilon - \frac{1}{\pi} \sin \pi\epsilon \right)$$

$$+ 2a_{21} \left[(1 - \epsilon) \cos \frac{\pi\epsilon}{2} + \frac{1}{2\pi} \left(\sin \frac{\pi\epsilon}{2} + \sin \frac{3\pi\epsilon}{2} \right) \right]$$

$$+ 2a_{22} \left[\frac{1}{\pi} \sin \pi\epsilon - \frac{1}{3\pi} (\sin \pi\epsilon - \sin 2\pi\epsilon) \right]$$

$$- a_{31} \left[(1 - \epsilon) \cos \pi\epsilon + \frac{1}{\pi} \sin \pi\epsilon \right]$$

$$\left. - a_{32} \left[\frac{1}{\pi} (\sin 2\pi\epsilon + \sin \pi\epsilon) - \frac{1}{3\pi} (\sin 2\pi\epsilon - \sin \pi\epsilon) \right] \right\}$$

$$(6.30)$$

where
$$\epsilon = \frac{2S \tan \phi}{L} \qquad (6.31)$$

In Eq. (6.30), the term within braces can be regarded as the difference between the first harmonic amplitude a_{11} actually occurring in girder 1 and the first harmonic amplitude A_{11} which would occur in the absence of transverse load distribution.

Second harmonic load and response of girder 1. The second harmonic deflection of girder 1 is $a_{12} \sin (2\pi x_1)/L$. This deflection requires a second harmonic load equal to $16k\, a_{12} \sin (2\pi x_1)/L$. The following equation can be written similarly to Eq. (6.29).

$$P_{L1}^{(2)} = 16k\, A_{12} \qquad (6.32)$$

where A_{12} is the amplitude of the second harmonic component of the free deflection of girder 1. Similarly to Eq. (6.28) we have

$$16k\, a_{12} = 16k\, A_{12} + \frac{2}{L} \int_{2S}^{L} Q_1 \sin \frac{2\pi x_1}{L} \, dx_1 \qquad (6.33)$$

This equation yields the following equation for a_{12}:

$$a_{12} = A_{12} + \frac{\eta}{128} \left\{ -a_{11} \left(\frac{1}{3\pi} \sin 3\pi\epsilon - \frac{1}{\pi} \sin \pi\epsilon \right) \right.$$

$$- a_{12} \left[(1 - \epsilon) + \frac{1}{4\pi} \sin 4\pi\epsilon \right]$$

$$+ a_{21} \frac{2}{3\pi} \left(-4 \sin \frac{\pi\epsilon}{2} - 3 \sin \frac{3\pi\epsilon}{2} + \sin \frac{5\pi\epsilon}{2} \right)$$

$$+ 2a_{22} \left[(1 - \epsilon) \cos \pi\epsilon + \frac{1}{4\pi} (\sin \pi\epsilon + \sin 3\pi\epsilon) \right]$$

$$+ a_{31} \frac{1}{3\pi} (4 \sin \pi\epsilon + 2 \sin 2\pi\epsilon)$$

$$- a_{32} \left[(1 - \epsilon) \cos 2\pi\epsilon + \frac{1}{2\pi} \sin 2\pi\epsilon \right] \Bigg\} \tag{6.34}$$

Complete set of equations. Equations (6.30) and (6.34) are the first two of a series of six equations for six unknowns a_{11}, a_{12}, a_{21}, a_{22}, a_{31}, and a_{32}. The remaining four equations are obtained similarly from the consideration of first and second harmonic relationships for girders 2 and 3. The set of six equations can be written in the following matrix notation

$$[G]\{a\} = \{R\} \tag{6.35}$$

The various terms of matrix $[G]$ and vectors $\{a\}$ and $\{R\}$ are defined in Fig. 6.14a, b, and c respectively. The terms shown in these figures make use, for brevity, of the following definition.

$$r_n = \frac{1}{n\pi} \sin n\pi\epsilon \tag{6.36}$$

It is instructive to study the 6×6 $[G]$ matrix shown in Fig. 6.14a. This matrix, which is formed entirely by two nondimensional parameters η and ϵ, can be conveniently regarded as being built up from four submatrices, each 3×3 in the following way:

$$[G] = \left[\begin{array}{c|c} G_{11} & G_{12} \\ \hline G_{21} & G_{22} \end{array} \right] \tag{6.37}$$

The submatrix $[G_{11}]$ relates the first harmonic responses of deflection to the first harmonic components of the external loads. Similarly, $[G_{12}]$ relates the second harmonic responses to the first harmonic component of the external loads, $[G_{21}]$ relates the first harmonic responses to the second harmonic components of the external loads, and $[G_{22}]$ relates the second harmonic responses to the second harmonic components of the external loads.

$$
\begin{bmatrix}
\dfrac{8}{\eta}+(1-\epsilon)+r_2 & -2(1-\epsilon+r_1)\cos\dfrac{\pi\epsilon}{2} & (1-\epsilon)\cos\pi\epsilon+r_1 & -(r_1-r_3) & -\dfrac{4}{3}(r_1+r_2) & \dfrac{4}{3}(r_1+r_2) \\[2mm]
-(1-\epsilon+r_1)\cos\dfrac{\pi\epsilon}{2} & \dfrac{4}{\eta}+2(1-\epsilon+r_1) & -(1-\epsilon+r_1)\cos\dfrac{\pi\epsilon}{2} & \dfrac{4}{3}(2r_1-r_2)\cos\dfrac{\pi\epsilon}{2} & 0 & -\dfrac{4}{3}(2r_1-r_2)\cos\dfrac{\pi\epsilon}{2} \\[2mm]
(1-\epsilon)\cos\pi\epsilon+r_1 & -2(1-\epsilon+r_1)\cos\dfrac{\pi\epsilon}{2} & \dfrac{8}{\eta}+(1-\epsilon)+r_2 & -\dfrac{4}{3}(r_1+r_2) & \dfrac{4}{3}(r_1+r_2) & r_1-r_3 \\[2mm]
-(r_1-r_3) & \dfrac{8}{3}(2r_1-r_2)\cos\dfrac{\pi\epsilon}{2} & -\dfrac{4}{3}(r_1+r_2) & \dfrac{128}{\eta}+(1-\epsilon)+r_4 & -2(1-\epsilon+r_2)\cos\pi & (1-\epsilon)\cos 2\pi\epsilon+r_2 \\[2mm]
-\dfrac{2}{3}(r_1+r_2) & 0 & \dfrac{2}{3}(r_1+r_2) & -(1-\epsilon+r_2)\cos\pi\epsilon & \dfrac{64}{\eta}+2(1-\epsilon+r_2) & -(1-\epsilon+r_2)\cos\pi\epsilon \\[2mm]
\dfrac{4}{3}(r_1+r_2) & -\dfrac{8}{3}(2r_1-r_2)\cos\dfrac{\pi\epsilon}{2} & r_1-r_3 & (1-\epsilon)\cos 2\pi\epsilon+r_2 & -2(1-\epsilon+r_2)\cos\pi\epsilon & \dfrac{128}{\eta}+(1-\epsilon)+r_4
\end{bmatrix}
\begin{Bmatrix}
a_{11} \\[1mm] a_{21} \\[1mm] a_{31} \\[1mm] a_{12} \\[1mm] a_{22} \\[1mm] a_{32}
\end{Bmatrix}
=
\begin{Bmatrix}
\dfrac{8}{\eta}A_{11} \\[1mm] \dfrac{4}{\eta}A_{21} \\[1mm] \dfrac{8}{\eta}A_{31} \\[1mm] \dfrac{128}{\eta}A_{12} \\[1mm] \dfrac{64}{\eta}A_{22} \\[1mm] \dfrac{128}{\eta}A_{32}
\end{Bmatrix}
$$

$\qquad(a)\qquad\qquad\qquad(b)\qquad\qquad\qquad(c)$

Figure 6.14 Equation (6.35). (a) [G] matrix. (b) {a} vector. (c) {R} vector.

It is noted that the submatrices $[G_{11}]$ and $[G_{22}]$ are polar symmetric with respect to their center elements, while $[G_{12}]$ and $[G_{21}]$ are polar antisymmetric. These properties follow from the facts that the structure itself is polar symmetric with respect to the midspan of the middle girder and that the first and second harmonic shapes are respectively symmetric and antisymmetric with respect to the midspans of the girders.

In the particular case of a bridge with zero angle of skew, i.e., a right bridge, all elements of $[G_{12}]$ and $[G_{21}]$ become zero, and the first and second harmonic effects are completely uncoupled.

In the case of a bridge with very poor load distribution characteristics, the value of η becomes very small. It can be seen that for very small values of η the principal diagonal terms of $[G]$ become dominant with similar corresponding values in vector $\{R\}$. It is clear that as η tends to zero, a_{11} tends to A_{11}, a_{21} tends to A_{21}, and so on. This is entirely to be expected, as it means that when the load distribution characteristics of the bridge are very poor, most of the external loads are retained by the directly loaded girders.

Equations for moments. Equation (6.35) is expressed in terms of deflection amplitudes. The conversion of this equation to correspond to moment amplitudes is readily accomplished, as shown in the following.

A deflection $a_{11} \sin (\pi x_1/L) + a_{12} \sin (2\pi x_1/L)$ corresponds to a bending moment $(\pi^2 EI/L^2)[a_{11} \sin (\pi x_1/L) + 4a_{12} \sin (2\pi x_1/L)]$, which is conveniently written as $[m_{11} \sin (\pi x_1/L) + m_{12} \sin (2\pi x_1/L)]$, so that

$$a_{11} = m_{11} \frac{L^2}{\pi^2 EI}$$

$$a_{12} = m_{12} \frac{L^2}{4\pi^2 EI}$$

$$a_{21} = m_{21} \frac{L^2}{\pi^2 EI}$$

$$a_{22} = m_{22} \frac{L^2}{4\pi^2 EI}$$

$$a_{31} = m_{31} \frac{L^2}{\pi^2 EI}$$

$$a_{32} = m_{32} \frac{L^2}{4\pi^2 EI}$$

(6.38)

On canceling a common factor $L^2/(\pi^2 EI)$ in vectors $\{a\}$ and $\{R\}$ on both sides of Eq. (6.35), the following set of equations is obtained.

$$
\begin{bmatrix} G \\ \text{matrix} \end{bmatrix}
\begin{Bmatrix}
m_{11} \\
m_{21} \\
m_{31} \\
\dfrac{1}{4}\,m_{12} \\
\dfrac{1}{4}\,m_{22} \\
\dfrac{1}{4}\,m_{32}
\end{Bmatrix}
=
\begin{Bmatrix}
\dfrac{8M_{L1}^{(1)}}{\eta} \\
\dfrac{4M_{L2}^{(1)}}{\eta} \\
\dfrac{8M_{L3}^{(1)}}{\eta} \\
\dfrac{32M_{L1}^{(2)}}{\eta} \\
\dfrac{16M_{L2}^{(2)}}{\eta} \\
\dfrac{32M_{L3}^{(2)}}{\eta}
\end{Bmatrix}
\tag{6.39}
$$

It is noted that the $[G]$ matrix is the same as given in Fig. 6.14a and that $M_{Li}^{(1)}$ and $M_{Li}^{(2)}$ are the first two harmonic components of free moments in girder i as defined by Eq. (6.9).

As can be seen from Eq. (6.9), the free moment M_{Li} in girder i is composed of an infinite number of harmonic components. Only the first two of these components are considered in the load distribution analysis, the remaining components being assumed to be retained entirely by the girder itself. Girder i retains $m_{i1} \sin{(\pi x_1/L)}$ out of the first component of free moment and $m_{i2} \sin{(2\pi x_i/L)}$ out of the second. Hence, the net moments M_i in girder i are given by

$$
M_i = m_{i1} \sin \frac{\pi x_i}{L} + m_{i2} \sin \frac{2\pi x_i}{L}
$$

$$
+ M_{Li}^{(3)} \sin \frac{3\pi x_i}{L} + M_{Li}^{(4)} \sin \frac{4\pi x_i}{L} + \cdots \tag{6.40}
$$

If the girder under consideration does not have any externally applied load on it, then clearly the third and higher terms on the right-hand side of Eq. (6.40) become equal to zero.

To obtain a reasonably accurate assessment of net moments M_i from Eq. (6.40), one must include in the calculations at least 10 terms of the free-moment series. Alternatively, and much more accurately, the convergence can be hastened by taking advantage of the fact that the sum of the third and all higher terms of the right-hand side of Eq. (6.9) is precisely equal to M_{Li} minus the moments due to the first two harmonics. Thus by using Eq. (6.9), Eq. (6.40) can be rewritten as

$$M_i = M_{Li} + (m_{i1} - M_{Li}^{(1)}) \sin \frac{\pi x_i}{L} + (m_{i2} - M_{Li}^{(2)}) \sin \frac{2\pi x_i}{L} \quad (6.41)$$

Physical aspects of analysis. The physical significance of the basis of analysis of three-girder skew bridges can be conveniently explained with reference to Fig. 6.15, which shows a three-girder skew bridge carrying a set of external loads on girder 3. As shown in Fig. 6.15b, this applied loading is represented for distribution effects by only the first and second harmonics, with the former being in the shape of one half sine wave and the latter in the form of two half sine waves, one of which acts downward and the other upward.

Now consider the transfer of the first harmonic component of the external loads to girder 2. If this girder were directly opposite girder 3 and the transverse medium extended the full length of both girders, the load transferred to both girders would also have been in the shape of a single half sine wave. For the case under consideration, however, the loading received by girder 2 is predominantly in the shape of a truncated half sine wave. Also, as shown in Fig. 6.15c, the loading in the shape of a truncated half sine wave can itself be conveniently represented by two terms of another harmonic series. Thus the first harmonic component of the external loads gets transferred to other girders not as purely first harmonic but also with some second harmonic content.

Figure 6.15d shows a similar procedure for the transfer of the second harmonic component of external loads, which also gets transferred to adjacent girders as both first and second harmonics.

All the relevant bridge properties are represented in the two dimensionless parameters η and ϵ in the analysis. Accordingly, these two are the characterizing parameters for load distribution in skew bridges with three girders. The significance and meaning of the term *characterizing parameters* is that the load distribution characteristics of two dissimilar-looking bridges are the same if they have the same values for these parameters.

To demonstrate the validity of the characterizing parameters η and ϵ, two different bridges having the same values for these parameters were analyzed by the grillage analogy method (e.g., see Ref. 3) for single concentrated loads on an outer girder. To facilitate a direct comparison of results, the skew spans were kept the same at 20.0 m for the two bridges. As shown in Fig. 6.16, one bridge had a skew angle of 45.00° and girder spacing of 2.5 m, and the other had a skew angle of 59.04° and girder spacing of 1.5 m. This led to η of 4.5 and ϵ of 0.25. Girders of both bridges had the same EI, but the values of D_y were adjusted to get the same value of η.

Moments in different girders of these two bridges, as obtained by the grillage analogy method, are plotted in Fig. 6.16 for two different load cases. It can be seen that the corresponding moment diagrams for the two bridges are imperceptibly different from each other, thus providing

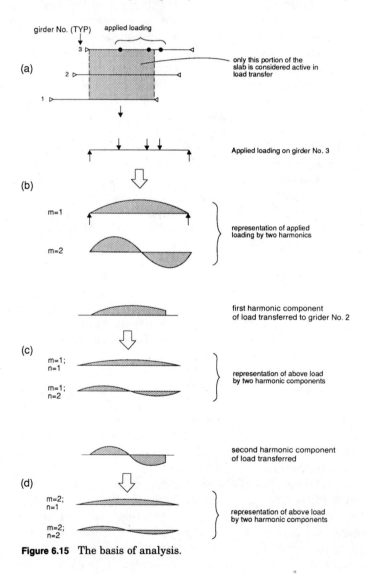

Figure 6.15 The basis of analysis.

a tangible demonstration that η and ε do indeed characterize the load distribution characteristics of skew slab-on-girder bridges.

It is instructive to note that the skew angle affects only one characterizing parameter, ε. The composition of this parameter also includes girder spacing and span, indicating that, contrary to the prevalent belief, the skew angle is not on its own the bridge parameter which distinguishes its load-carrying characteristics from those of the corresponding right bridge.

Figure 6.17a and b shows single and double half sine waves, re-

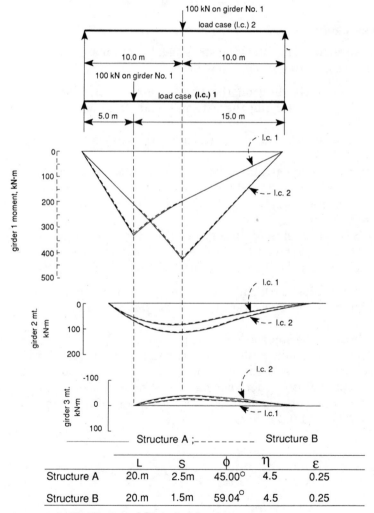

	L	S	φ	η	ε
Structure A	20.m	2.5m	45.00°	4.5	0.25
Structure B	20.m	1.5m	59.04°	4.5	0.25

Figure 6.16 Verification of characterizing parameters.

spectively plotted over the same length, and Fig. 6.17c shows their sum. It is interesting to note in Fig. 6.16 that the moment diagrams for girders 2 and 3, which do not carry any external loads, have the same general shape as the curve of Fig. 6.17c. In other words, the moment diagrams for these girders are predominantly composed of the first two harmonics, as was assumed for the semicontinuum method. Since the analysis whose results led to these moment diagrams was effected without incorporating any harmonic series, it is confirmed that the composition of the moments as being the sum of two harmonics is a reality and not the outcome of the assumptions of analysis.

6.3.3   Application of analysis

Steps of calculation.   The following steps of calculation are required for analyzing a given three-girder skew bridge.

1. From applied loading on the various girders, obtain $M_{Li}^{(1)}$ and $M_{Li}^{(2)}$ for each girder carrying the external loads from equations such as Eq. (6.10).

2. Using Eqs. (6.26) and (6.31), calculate the values of η and ϵ respectively.

3. Using the above calculated values of η and ϵ, construct the $[G]$ matrix shown in Fig. 6.14a, where r_n is as defined by Eq. (6.36).

4. Solve the set of equations defined by Eq. (6.39) to obtain m_{i1} and m_{i2} for $i = 1, 2,$ and 3.

5. Obtain net moments M_i in a girder i from Eq. (6.41), where M_{Li} is the free moment in the girder obtained by elementary beam analysis at a distance x_i from the left-hand reference support of the girder.

Clearly these steps are ideally suited to being carried out on a personal computer, using a program written for that purpose.

Worked example.   The application of the proposed analysis is illustrated with the help of the following worked example for the bridge whose details are given in Fig. 6.18. As shown in this figure, the bridge has a skew span of 30.0 m, girder spacing of 2.54 m, and a skew angle of 60°. From these and other relevant properties of the bridge η and ϵ are found to be 10.0 and 0.3 respectively.

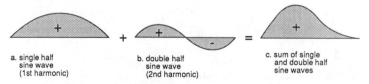

a. single half
sine wave
(1st harmonic)

b. double half
sine wave
(2nd harmonic)

c. sum of single
and double half
sine waves

Figure 6.17 Sum of two harmonics.

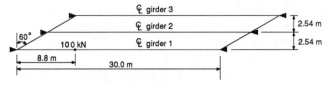

Figure 6.18 Plans of a skew bridge.

As shown in Fig. 6.18, girder 1 carries a point load of 100 kN at 8.8 m from its left-hand support. It is required to calculate the moments in the externally loaded girder.

From Eq. (6.10), $M_{L1}^{(1)}$ and $M_{L1}^{(2)}$ are found to be 484.2 and 146.4 kN·m respectively, the $[G]$ matrix is constructed for η and ϵ of 10.0 and 0.3 respectively, and Eq. (6.39) is solved to give m_{11} and m_{12} of 383.3 and 140.5 kN·m respectively. The net moment M_1 in girder 1 is calculated for six different values of x_1 by using the already obtained values of $M_{L1}^{(1)}$, $M_{L1}^{(2)}$, m_{11}, and m_{12} and the free moment M_{L1}. The values of x_1 and M_{L1} are listed in Table 6.1 for the six reference stations together with the other steps of calculation which lead to final values of moments in girder 1.

Comparison with grillage analysis. The moments M_1 obtained above for girder 1 are plotted along the girder length in Fig. 6.19 along with the girder 1 moments obtained by grillage analysis. This figure also shows the idealizations used in the semicontinuum and grillage analyses.

As can be seen in Fig. 6.19, the correlation between the results of the two methods is excellent, with the moments obtained by the semicontinuum method being larger by a very small margin. These and other comparisons, some of which are given in Ref. 4, have verified that the semicontinuum method for three-girder skew bridges is reliable and predicts results similar to those obtained by grillage analysis.

6.3.4 Manual method

Since the number of unknowns in a three-girder skew bridge is not unduly large, it is feasible to apply the semicontinuum method of Sec. 6.3.2 manually with the help of charts given in the section.

TABLE 6.1 Calculation of Moments in Girder 1 of the Bridge Shown in Fig. 6.18

x_1, m	4.40	8.80	13.04	17.28	21.52	25.76
M_{L1}, kN·m	310.9	621.9	497.5	373.1	248.5	124.4
$(383.3 - 484.2) \sin(\pi x_1/30)$, kN·m	−44.9	−80.4	−98.8	−98.1	−78.4	−43.4
$(140.5 - 146.4) \sin(2\pi x_1/30)$, kN·m	−4.7	−5.6	−2.3	2.7	5.7	4.5
M_1, kN·m	261.3	535.9	396.4	277.7	175.8	85.5

Similarly to the method for right bridges given in Chap. 10, use is made of distribution coefficients defined in Chap. 1. Unlike the manual method of Chap. 10, however, two harmonic terms of applied loading are considered in the analysis, each of which results in transfer to the other girders in both first and second harmonics. It follows, therefore, that four different distribution coefficients, corresponding to each loading condition, must be considered for each girder. Accordingly, a four-subscripted notation is introduced for the distribution coefficient ρ.

A distribution coefficient for the three-girder skew bridge is denoted as $(\rho_{i,j})_{m,n}$

where i = number of the girder whose response is investigated
   j = number of the externally loaded girder
   m = harmonic number of the external-load component
   n = harmonic number of the transferred-load component

Thus, for example, $(\rho_{2,3})_{2,1}$ is the distribution coefficient for girder 2 when the external load is on girder 3 and when the second harmonic component of external load is causing a first harmonic response in the girder under consideration. This notation will be clearly recognizable

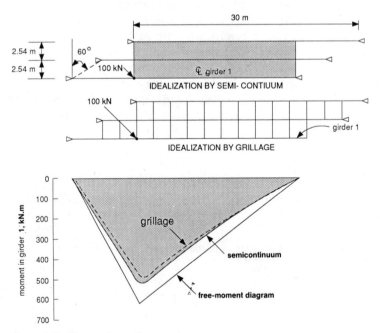

Figure 6.19 Comparison of moments.

from the m,n labels attached to the various portions of Fig. 6.15b, c, and d.

From the method of Sec. 6.3.2, it can be shown that the moment M_i in girder i due to external loads on girder j can be obtained from the distribution coefficients by the following expression.

$$M_i = \left[(\rho_{i,j})_{1,1} M_{Lj}^{(1)} + \frac{1}{4} (\rho_{i,j})_{2,1} M_{Lj}^{(2)} \right] \sin \frac{\pi x_i}{L}$$

$$+ [4(\rho_{i,j})_{1,2} M_{Lj}^{(1)} + (\rho_{i,j})_{2,2} M_{Lj}^{(2)}] \sin \frac{2\pi x_i}{L} \quad (6.42)$$

where $M_{Lj}^{(1)}$ and $M_{Lj}^{(2)}$ are obtained from Eq. (6.10) or similar equations.

The moments in a directly loaded girder, i.e., in girder i, due to load also on girder i are obtained by subtracting from the free moments those moments which are passed onto the other girders. By using this criterion the expression for moments M_i in the externally loaded girder i becomes

$$M_i = M_{Li} + \left\{ [(\rho_{i,i})_{1,1} - 1] M_{Li}^{(1)} + \frac{1}{4} (\rho_{i,i})_{2,1} M_{Li}^{(2)} \right\} \sin \frac{\pi x_i}{L}$$

$$+ \{4(\rho_{i,i})_{1,2} M_{Li}^{(1)} + [(\rho_{i,i})_{2,2} - 1] M_{Li}^{(2)}\} \sin \frac{2\pi x_i}{L} \quad (6.43)$$

Values of the various distribution coefficients were calculated by the method of Sec. 6.3.2 for pairs of values of η and ϵ covering the entire practical range of three-girder skew bridges. These coefficients are plotted in Fig. 6.20a through f in the form of contours. By using these charts three-girder skew bridges can be analyzed rigorously by manual calculations. The steps of calculation remain the same as given in Sec. 6.3.3 except that steps 3 and 4 are replaced by the reading of the values of the distribution coefficients from Fig. 6.20 and that the girder moments are obtained by the relevant one of Eqs. (6.42) and (6.43).

Worked example. In this example, a 45° skew bridge is analyzed by the manual method for a loading which comprises two patches of uniformly distributed loads. This is a bridge design loading often used on the Indian subcontinent and referred to as the 70-R loading. The plan of the bridge and load position on the plan are shown in Fig. 6.21. The moduli of elasticity of the girder and deck-slab material are assumed to be the same. The moment of inertia I of a girder is 0.22 m⁴. For a slab thickness of 0.22 m, $D_y = E \times 8.87 \times 10^{-4}$. Hence, from

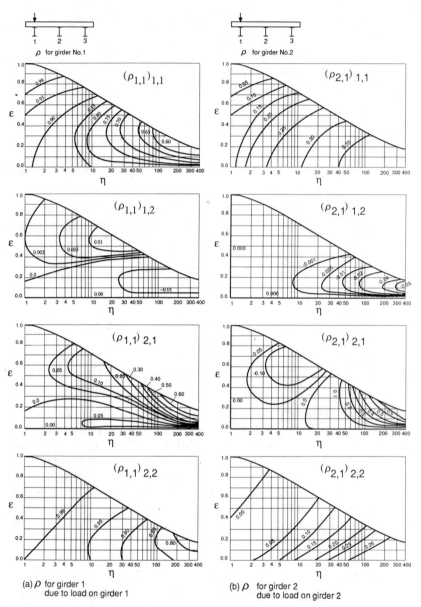

(a) ρ for girder 1
due to load on girder 1

(b) ρ for girder 2
due to load on girder 2

Figure 6.20 Distribution coefficients.

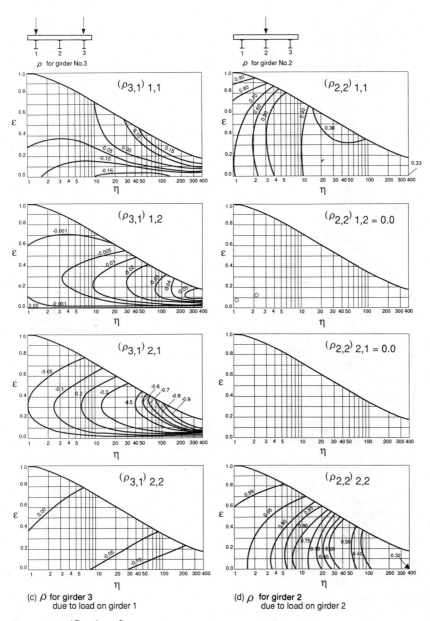

(c) ρ for girder 3
    due to load on girder 1

(d) ρ for girder 2
    due to load on girder 2

Figure 6.20 *(Continued)*

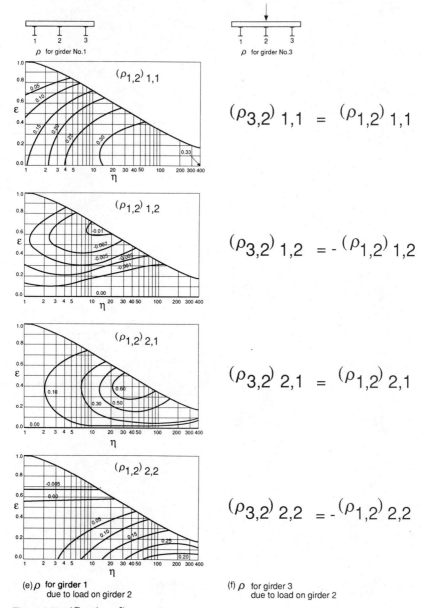

$$(\rho_{3,2})_{1,1} = (\rho_{1,2})_{1,1}$$

$$(\rho_{3,2})_{1,2} = -(\rho_{1,2})_{1,2}$$

$$(\rho_{3,2})_{2,1} = (\rho_{1,2})_{2,1}$$

$$(\rho_{3,2})_{2,2} = -(\rho_{1,2})_{2,2}$$

(e) ρ for girder 1
due to load on girder 2

(f) ρ for girder 3
due to load on girder 2

Figure 6.20 (*Continued*)

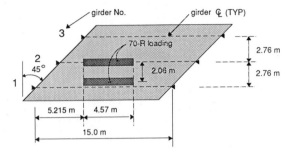

Figure 6.21 Plan of a skew bridge.

Eqs. (6.26) and (6.31)

$$\eta = \frac{12}{\pi^4}\left(\frac{15.0}{2.76}\right)^3 \frac{15 \times E \times 8.87 \times 10^{-4}}{E \times 0.22} = 1.20$$

$$\epsilon = \frac{2 \times 2.76 \tan 45°}{15} = 0.37$$

The relevant distribution coefficients corresponding to $\eta = 1.20$ and $\epsilon = 0.37$ are read from the charts of Fig. 6.20 and are listed in Table 6.2.

As shown in Fig. 6.22a, the centerline of one line of tracks, weighing 35 t, of the design vehicle is directly above girder 2 and the other line of tracks is between girders 1 and 2 at a distance of 0.79 m from the axis of the former girder. From statics, the second line of tracks is distributed between girders 1 and 2, with the portion going to girder 2 being $35 \times 0.79/2.76 \approx 10.0$ t. Accordingly, as shown in Fig. 6.22b, the total load apportioned to girder 1 is 25 t and that to girder 2 is 45 t.

By using App. I, the free moment M_{Li} on a girder i due to a uniformly

TABLE 6.2 Distribution Coefficients for $\eta = 1.20$ and $\epsilon = 0.37$

Load on girder i	Distribution coefficient for girder i	$(\rho_{i,j})_{1,1}$	$(\rho_{i,j})_{1,2}$	$(\rho_{i,j})_{2,1}$	$(\rho_{i,j})_{2,2}$
1	1	0.93	0.00	0.03	1.00
	2	0.12	0.00	−0.02	0.00
	3	−0.05	0.00	−0.05	0.00
2	1	0.12	0.00	0.05	0.01
	2	0.70	0.00	0.00	0.97
	3	0.12	0.00	−0.05	−0.01

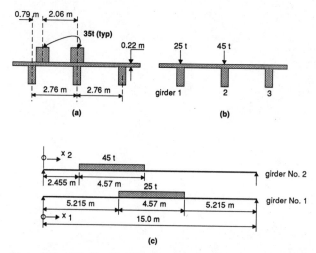

Figure 6.22 Details of loading for the worked example. (a) Actual transverse vehicle position. (b) Transformed loads on girders. (c) Longitudinal positions of loads on girders 1 and 2.

distributed load W of length $2u$ can be shown to be given by the following expression.

$$M_{Li} = \frac{2WL^2}{\pi^3 u} \sum_{n=1}^{\infty} \frac{1}{n^3} \sin \frac{n\pi c}{L} \sin \frac{n\pi u}{L} \sin \frac{n\pi x}{L} \qquad (6.44)$$

TABLE 6.3 Calculation of Moments in the Bridge Shown in Figs. 6.17 and 6.18

Row no.	Girder no.	x	m				
			2.5	5.0	7.5	10.0	12.5
1	1	Free moment, M_{L1}	31.25	62.5	79.47	62.50	31.25
2		2d term of Eq. (6.43)	−2.55	−4.43	−5.11	−4.43	−2.55
3		3d term of Eq. (6.43)	0.00	0.00	0.00	0.00	0.00
4		1st term of Eq. (6.42)	6.81	11.79	13.61	11.79	6.81
5		2d term of Eq. (6.42)	0.26	0.26	0.00	−0.26	−0.26
6		Σ	35.77	70.12	87.97	69.60	35.25
7	2	Free moment, M_{L2}	76.61	121.91	106.65	71.10	35.55
8		2d term of Eq. (6.43)	−16.50	−28.64	−33.08	−28.64	−16.50
9		3d term of Eq. (6.43)	−0.80	−0.80	0.00	0.80	0.80
10		1st term of Eq. (6.42)	4.39	7.60	8.77	7.60	4.39
11		2d term of Eq. (6.42)	0.00	0.00	0.00	0.00	0.00
12		Σ	63.70	100.07	82.34	50.86	24.24
13	3	1st term of Eq. (6.42)	−1.83	−3.17	−3.66	−3.17	−1.83
14		2d term of Eq. (6.42)	0.00	0.00	0.00	0.00	0.00
15		1st term of Eq. (6.42)	6.80	11.79	13.61	11.79	6.80
16		2d term of Eq. (6.42)	−0.15	−0.15	0.00	0.15	0.15
17		Σ	4.82	8.47	9.95	8.77	5.12

where c is the distance of the middle of the load from the reference support and the length $2u$ of the load is smaller than L.

The longitudinal positions and lengths of the apportioned loads on girders 1 and 2 are shown in Fig. 6.22c. By substituting the values of u, c, w, and L in Eq. (6.44) values are obtained for the two harmonic components for girder 1:

$$M_{L1}^{(1)} = 73.12 \text{ t} \cdot \text{m} \qquad M_{L1}^{(2)} = 0$$

Similarly, for girder 2,

$$M_{L2}^{(1)} = 110.23 \text{ t} \cdot \text{m} \qquad M_{L2}^{(2)} = 30.79 \text{ t} \cdot \text{m}$$

The various steps of calculating the net girder moments for different values of x are listed in Table 6.3. Some of these steps are discussed as follows with reference to the row numbers of this table.

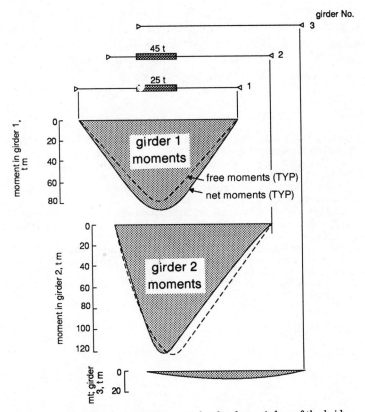

Figure 6.23 Bending-moment diagrams for the three girders of the bridge shown in Figs. 6.21 and 6.22.

Row 1 contains the free moments in girder 1 due to the 25-t uniformly distributed load. For $i = 1$, the second and third terms within brackets of the right-hand side of Eq. (6.43) correspond to the first and second harmonic moments, respectively, that are passed to girders 2 and 3; these moments are given in rows 2 and 3. The moments in rows 4 and 5 are the first and second harmonic moments, respectively, that are transferred to girder 1 owing to loading on girder 2. The sum of moments in rows 1 through 5 are the net moments in girder 1 and are given in row 6.

A similar procedure is followed in rows 7 through 11 for obtaining the net moments in girder 2, which are given in row 12.

Since girder 3 does not have any externally applied loads, the net moments in it are only those which are passed on from girders 1 and 2. Moments due to the former are given in rows 13 and 14, and those due to the latter are given in rows 15 and 16.

The diagrams for net bending moments in girders 1 and 2 are compared in Fig. 6.23 with the free-moment diagrams. The figure also contains the moment diagram for girder 3. It can be seen that the moments in girder 3 are very small compared with those in the externally loaded girders. This behavior indicates that the bridge under consideration, which is not unusual in some parts of the world, has extremely poor load distribution characteristics.

REFERENCES

1. American Association of State Highway and Transportation Officials (AASHTO): *Standard Specifications for Highway Bridges,* Washington, 1983.
2. Bakht, B., and Moses, F.: Lateral distribution factors for highway bridges, *ASCE Journal of Structural Division,* 1988.
3. Jaeger, L. G., and Bakht, B.: The grillage analogy in bridge analysis, *Canadian Journal of Civil Engineering,* 9(2), 1982, pp. 224–235.
4. Jaeger, L. G., Bakht, B., and Surana, C. S.: Application of analysis of three-girder skew bridges. *Proceedings, Second International Colloquium on Concrete in Developing Countries, Bombay,* January 1988, sec. 5, pp. 52–66.
5. Surana, C. S.: Generalised analysis of interconnected skew girder bridges and grids, *Proceedings, Twenty-second Congress of the Indian Society of Theoretical and Applied Mechanics,* December 1977, pp. 7–17.
6. Surana, C. S.: Load distribution in skew-span girder bridge decks, *Bridge and Structural Engineer, India,* 8(1), 1978, pp. 13–23.

Slab and Similar Bridges

As discussed in Sec. 1.2.3, a slab bridge can be analyzed by the grillage analogy method (e.g., see Ref. 3). For this method the Poisson's ratio of the slab material is usually neglected, and the slab is idealized as an assembly of orthogonal beams. Figure 7.1b shows the grillage idealization of the slab bridge shown in Fig. 7.1a. Clearly, if a slab bridge can be idealized as an assembly of discrete beams, it can also be idealized as a semicontinuum. In fact, the latter idealization, an example of which is shown in Fig. 7.1c, is a closer representation of the slab in the transverse direction.

This chapter deals with the analysis of right slab bridges and certain voided-slab bridges by the semicontinuum method of Chap. 4 and provides guidance on the interpretation of the results of such analysis. Cross sections of slab and voided-slab bridges are shown in Fig. 1.13a and b respectively.

7.1 Idealization

7.1.1 Slab bridges

For the semicontinuum idealization, a slab bridge can be conceptually divided into any number of longitudinal strips, with each strip represented by a longitudinal beam of the idealization. These conceptual

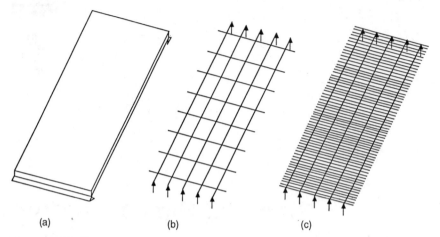

Figure 7.1 Idealizations of a slab bridge. (*a*) Slab bridge. (*b*) Idealization by a grillage. (*c*) Idealization by a semicontinuum.

beams can have different widths; however, it is advisable to keep their widths equal except for the two outer strips, whose widths may suitably be one-half of those of the inner ones. The outer and inner strips are identified in Fig. 7.2, which also shows examples of the two preferred idealizations.

When a bridge is conceptually divided into strips of equal widths, as shown in Fig. 7.2*a*, each longitudinal beam of the idealization is placed at the middle of the strip it represents. In the other case, which is shown in Fig. 7.2*b*, the inner beams lie at the middle of the strips they represent, but the outer ones lie at the outer boundaries of the corresponding strips. It can be readily shown that the flexural rigidity *EI* of a beam representing a strip of width *S* is given by

$$EI = SD_x \qquad (7.1)$$

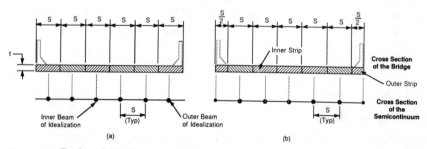

Figure 7.2 Preferred idealizations. (*a*) Division by equally spaced strips. (*b*) Division with outer strips at half spacing.

where D_x is the longitudinal flexural rigidity per unit width of the bridge. For a slab of thickness t, it is given by

$$D_x = \frac{Et^3}{12(1 - v^2)} \qquad (7.2a)$$

Since the Poisson's ratio v of concrete is small, being equal to about 0.15, the term $(1 - v^2)$ is very nearly equal to 1.0. Accordingly, D_x can be obtained from the following expression without incurring any significant error in the load distribution analysis.

$$D_x = \frac{Et^3}{12} \qquad (7.2b)$$

The torsional rigidity GJ of a semicontinuum beam representing a strip of the slab of width S is given by

$$GJ = SD_{xy} \qquad (7.3)$$

where D_{xy}, the longitudinal torsional rigidity of the slab per unit width, can be obtained from

$$D_{xy} = \frac{Gt^3}{6} \qquad (7.4)$$

The torsional rigidity of a rectangular prismatic bar of width W and thickness t, shown in Fig. 7.3, is equal to $GWt^3/3$ provided that t is very small in comparison with W. This gives the torsional rigidity per unit width of the bar to be equal to $Gt^3/3$, which is twice as much as D_{xy}. The apparent discrepancy between the torsional rigidities of the slab and a bar can be resolved readily by noting that the torsional rigidity of an isotropic slab is divided into two equal and complementary rigidities in longitudinal and transverse directions, namely, D_{xy} and D_{yx} respectively. The torsional rigidity GJ of the bar shown in Fig. 7.3 is obtained by multiplying the sum of D_{xy} and D_{yx} by W; it can be seen that the value of GJ thus obtained is equal to $Gt^3/3$.

When a slab is represented only by a simple beam element, as it is for the analysis of the bar shown in Fig. 7.3, then the only value of the torsional rigidity used in the analysis is GJ, whose value is equal to $GWt^3/3$. However, when the slab is represented by a set of orthogonal beams, then the torsional rigidity is divided in equal intensities in the two orthogonal directions.

The slab itself constitutes the transverse medium. Hence its thickness, modulus of elasticity, and shear modulus can be used directly to represent the transverse medium.

7.1.2 Voided-slab bridges

Concrete slab bridges with circular or nearly circular voids running longitudinally are usually referred to as *voided-slab bridges*. When the voids are large compared with the slab thickness, the distortion of the cross section under nonuniform loads becomes so significant that it cannot be neglected. For such cases the so-called shear-weak semicontinuum method, in which the transverse continuum is assumed to have a finite, equivalent shear rigidity, can be used. However, when the voids are small, the semicontinuum method of Chap. 4 can be used in the same manner as for solid-slab bridges. Reference 5 recommends that the effects of the distortion of the cross section can be safely neglected when both of the following two conditions are satisfied.

1. The void diameter is less than 80 percent of the total slab thickness.

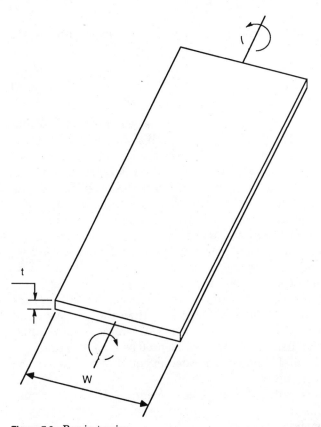

Figure 7.3 Bar in torsion.

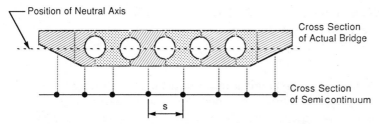

Figure 7.4 Idealization of a voided-slab bridge by a semicontinuum.

2. The center-to-center spacing of voids is more than the total slab thickness.

The cross section of a typical voided-slab bridge is shown in Fig. 7.4 together with the cross section of the recommended semicontinuum idealization. It can be seen in this figure that the slab in the middle portion is divided into segments by vertical planes passing the centers of the voids. In the outer portions, the segments are formed by dividing the slab in such a way that the flexural rigidities of all segments are nearly the same. Each conceptual segment is then represented by a longitudinal beam located at the transverse position of its center of gravity. The flexural rigidity of each segment is recommended to be calculated about a common neutral axis, which is straight, as shown in Fig. 7.4; the position of the neutral axis is obtained by considering the whole cross section.

As recommended in Ref. 2, the longitudinal torsional rigidity D_{xy} and the transverse flexural rigidity D_y of a voided slab with circular voids can be conveniently obtained by the following expressions, where the notation is as defined in Fig. 7.5.

$$D_{xy} = \frac{Gt^3}{6} \left[1 - 0.85 \left(\frac{t_v}{t} \right)^4 \right] \tag{7.5}$$

$$D_y = \frac{Et^3}{12} \left[1 - \left(\frac{t_v}{t} \right)^4 \right] \tag{7.6}$$

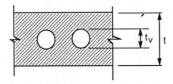

Figure 7.5 Partial cross section of a voided-slab bridge.

The torsional rigidity GJ of a longitudinal beam of the semicontinuum idealization is then given by

$$GJ = SD_{xy} \tag{7.7}$$

where S is spacing of the longitudinal beams. D_{yx}, the torsional rigidity per unit length of the transverse medium, has the same value as D_{xy}, which is given by Eq. (7.5).

The introduction of the values of D_y and D_{yx}, as obtained above, in the semicontinuum method poses no problem if these values are input directly. However, if a version of the method, e.g., SECAN1, which is developed specifically for slab-on-girder bridges, is used, it may require as input the thickness of the deck slab and the values of its E and G. In such a case, the following adjustments can be made to adapt the method to voided-slab bridges.

By using Eq. (7.6), the thickness t_c of the equivalent conceptual solid slab representing the transverse medium can be shown to be given by

$$t_c = t \left[1 - \left(\frac{t_v}{t} \right)^4 \right]^{0.33} \tag{7.8}$$

In order to represent the transverse torsional rigidity of the voided slab, a suitable adjustment should be made to the shear modulus of the transverse medium. By using Eqs. (7.5), (7.6), and (7.8), the expression for the equivalent shear modulus G_c of the transverse medium is found to be as follows:

$$G_c = G \, \frac{1 - 0.85 \left(\dfrac{t_v}{t} \right)^4}{1 - \left(\dfrac{t_v}{t} \right)^4} \tag{7.9}$$

7.2 Interpretation of Results

The semicontinuum method gives the intensities of transverse responses which can be directly interpreted as the corresponding responses of the slab or voided-slab bridge. For obtaining the intensities of longitudinal responses in the actual structure, however, the responses of longitudinal beams of the semicontinuum idealizations have to be divided by the respective widths of the slab that these beams represent. Other adjustments also need to be made to the results of the semicontinuum analysis as discussed in Secs. 7.2.1 and 7.2.2. The following notation is used for intensities of longitudinal responses.

M_x Intensity of longitudinal moment per unit width
V_x Intensity of longitudinal shear per unit width
M_{xy} Intensity of longitudinal twisting moment per unit width

7.2.1 Effect of Poisson's ratio

As pointed out in Ref. 3, a slab bridge, strictly speaking, can be idealized as an assembly of orthogonal beams only if the Poisson's ratio v of the slab material is assumed to be zero. By inference this restriction also applies to the idealization of slab and voided-slab bridges by the semicontinuum method.

The method for approximately taking account of the nonzero Poisson's ratio v on the various bridge responses is presented in Ref. 4. In this reference, it is shown that the method was developed by analyzing a large number of isotropic slabs of different aspect ratios for zero and nonzero values of v, using the isotropic-slab method of Ref. 1. Details of the slabs and loadings are given in Fig. 7.6, in which it can be seen that the slabs were analyzed for two different patch loads, each of them uniformly distributed. These loads, which were selected as representing vehicle loads, extend the full length of the slab, but in the transverse direction they occupy only a partial width.

The difference between responses, namely, moments and shear, corresponding to zero and nonzero values of v can be quantified by a factor F, which is defined by the following expression for transverse moments:

$$(M_y)_v = (M_y)_0 + F_{my} \times v \times (M_x)_0 \qquad (7.10)$$

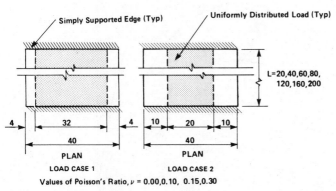

Figure 7.6 Details of bridges analyzed to determine effects of Poisson's ratio.

where M_x and M_y are intensities of longitudinal and transverse moments respectively and their subscripts refer to the values of Poisson's ratio to which they correspond. The subscript to F refers to the response to which the factor applies. Thus, for example, the factor for transverse moment intensity M_y is F_{my}, and that for the intensity of longitudinal shear V_x is F_{vx}.

The following conclusions were drawn from the results of the analyses reported in Ref. 4:

1. The value of F for a certain response at a given reference station is virtually the same for both load cases, thus confirming that, for all practical purposes, F is independent of the loading.

2. The effect of v on M_x and V_x is extremely small. The maximum effect is in the slab having a span-to-width ratio of 0.5 and v of 0.3. Even in this case, neglect of v changes M_x by less than 2 percent. No adjustment is therefore recommended to be made to the longitudinal moment and shear intensities obtained by the semicontinuum analysis.

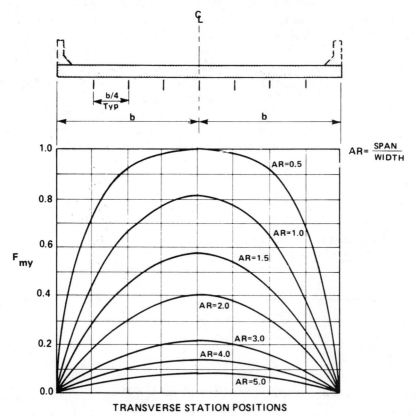

Figure 7.7 Correction factors for transverse moments.

3. The value of F_{my}, while being independent of the longitudinal position of the reference point, does vary across the transverse section with a zero value at the free edges and a maximum in the middle.

4. In general, the value of F_{my} decreases as ν increases. However, the variation is sufficiently small that F_{my} can be safely assumed to have the same value for all practical values of ν.

On the basis of the conclusions listed above, Ref. 4 recommends that corrections to account for the nonzero value of ν need to be made only for M_y; this can be done with the help of Eq. (7.10) and the charts for F_{my} given in Fig. 7.7. These charts were constructed from the results of isotropic-slab analyses mentioned earlier for a Poisson's ratio of 0.1. It is noted that for higher values of ν the factors given in these charts will lead to a slight overestimation of the transverse moments.

A feel for the effect of Poisson's ratio on transverse moments in a slab bridge can be developed readily with the help of a simple example, which is that of a rectangular slab simply supported along two opposite edges and subjected to some out-of-plane loads in its middle region, as shown in Fig. 7.8a. The cross section of a narrow longitudinal strip in the middle region of the slab is considered. The initially vertical edges of this strip are identified as AD and BC in Fig. 7.8b. It can be readily appreciated that if the strip were isolated from the rest of the slab, then owing to the Poisson's ratio of the slab material the bottom fibers of the strip would contract laterally and the top fibers expand. In the

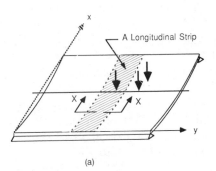

(a)

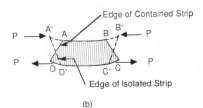

(b)

Figure 7.8 Effect of Poisson's ratio. (a) Rectangular plate in bending. (b) Cross section of a longitudinal strip at x–x.

cross section shown in Fig. 7.8b, the edges of the strip separated from the rest of the slab may assume positions $A'D'$ and $B'C'$. To bring these edges back to their original positions, i.e., AD and BC respectively, one would need a conceptual force P at each corner of the cross section acting as shown in Fig. 7.8b. These forces clearly induce a moment in the transverse direction. The transverse moment thus induced can be regarded as the contribution of the longitudinal moment in the transverse direction due to the effect of Poisson's ratio. The second term on the right-hand side of Eq. (7.10) is equivalent to this contribution, which is quantified by the factor F_{my}.

The magnitude of the force P, shown in Fig. 7.8b, clearly depends upon the containment of the strip under consideration by the expanse of the slab on either side of it. If the slab is long and narrow, the force P will be small. Similarly, if the strip is located near a free edge of the slab, this force will again be small. It is comforting to note that this pattern of behavior is confirmed by the charts given in Fig. 7.7. It can be seen that F_{my} varies from zero at the free edges to a maximum value at the middle; similarly, its value diminishes as one goes from short, wide bridges toward long, narrow ones.

While the charts presented in Fig. 7.7 were developed for solid-slab bridges, they can also be used safely for voided-slab bridges.

7.2.2 Spacing of longitudinal beams

A slab bridge can be idealized as a semicontinuum by using any number of longitudinal beams. At a cursory glance it seems intuitively obvious that a closer spacing of the longitudinal beams would lead to more accurate estimates of the peak intensities of moments and shears corresponding to these beams. In what follows in this section, it is shown that these peak intensities are not as sensitive to the spacings of beams as they are generally believed to be.

The usual practice of idealizing a slab bridge by a grillage and of interpreting the grillage analysis results is illustrated in Fig. 7.9 for a central concentrated load on a cross section of a slab bridge. As shown in this figure, the intensity of longitudinal moment in the slab at a longitudinal-beam location is obtained by dividing the beam moment by the width it represents. The transverse distribution curve of longitudinal-moment intensity is then obtained by joining the moment intensity points at the various beam locations.

The moment intensity curve of Fig. 7.9 is redrawn in Fig. 7.10 with emphasis on moments corresponding to the central beam, which is directly under the load. The average moment intensity for the region represented by this beam is shown by heavy dashed lines, and the continuous moment intensity curve obtained by the usual procedure

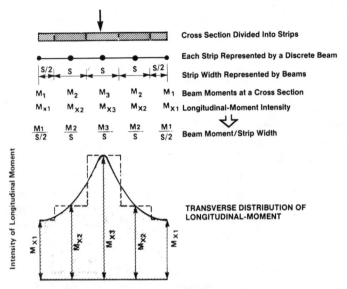

Figure 7.9 Idealization of a slab bridge by a grillage and interpretation of results.

by a heavy solid line. The beam moment and the area under the average intensity curve both represent integrated moment intensities over the width S represented by the beam. Accordingly, the total area under the continuous moment intensity curve over width S should be equal to the corresponding beam moment. As can be seen in Fig. 7.10a, this is not the case when the usual procedure is followed. The areas between

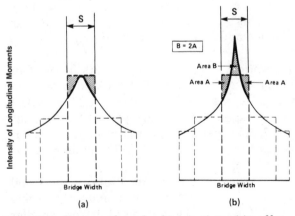

Figure 7.10 Two procedures for obtaining intensities of longitudinal moments. (a) Usual procedure. (b) Proposed procedure.

the average moments and those obtained by the usual procedure are shown as shaded in Fig. 7.10a and represent the degree of error in this procedure.

Graphical procedure for loads in the central region. It immediately follows that, to avoid this error, the curve of the interpreted moment intensities should be so drawn that the area under it within the beam spacing is equal to the corresponding beam moment. For the middle beam of Fig. 7.9, the moment intensity directly under the load should be larger than the average intensity so that, as shown in Fig. 7.10b, the area under the resulting curve is equal to the beam moment.

The results of semicontinuum analysis can be interpreted either graphically by the procedure illustrated in Fig. 7.10b or by the numerical techniques described later in this section. The use of the proposed graphical procedure is explained in the following paragraphs with the help of two specific examples.

A fictitious slab bridge having a span and width of 10 and 15 units respectively is analyzed by the semicontinuum method of Chap. 3 for a central-point load of 10 units. The bridge is analyzed for two different idealizations, in one of which it is divided conceptually into five longitudinal strips and in the other into four.

Midspan moments in the beams of the former idealization are given by the semicontinuum method to be 1.57, 6.37, 9.12, 6.37, and 1.57 units respectively. These moments lead to average moment intensities of 0.53, 2.12, 3.04, 2.12, and 0.53 units in the five strips respectively. The average moment intensities for the five-beam idealization are plotted in Fig. 7.11a together with the continuous moment intensity curve, which is plotted by using engineering judgment so that for each strip the area under the continuous curve remains the same as that under the average intensity line. This graphical interpretation gives a peak intensity of longitudinal moment under the load of approximately 3.15 units.

For the four-strip idealization, the midspan moments in the four longitudinal beams are found to be 2.65, 9.85, 9.85, and 2.65 units respectively, giving average intensities of 0.71, 2.63, 2.63, and 0.71 units respectively. These average intensities are plotted for the four-strip idealization in Fig. 7.11b together with the graphically interpreted continuous curve. In this case, the peak value is found to be 3.20 units.

It can be seen that for the two very close idealizations the usual practice gives peak intensities of longitudinal moments of 3.04 and 2.63 units respectively, with the latter being some 14 percent smaller than the former. The peak values of moments obtained by the graphical method, however, are within 3 percent of each other. It might perhaps be raised as an objection that because of relying on engineering judgment, the graphically interpreted values of peak intensities may vary more widely from engineer to engineer than for the examples given here. It may be noted that the dependence of interpreted values on idealization is reduced as one looks at responses integrated over finite widths rather than at peak intensities. Since loads themselves have finite widths, it seems appropriate that the former quantities be used in the design or evaluation process.

Fortunately, in real-life bridges there is always more than one concentrated load, and the moment and shear intensities near their respective peaks therefore do not change as rapidly as they do when there is only a single load. Consequently, in actual conditions the possibility of error involved in the proposed graphical process is further reduced to the point of being negligible.

Analytical procedure for eccentric loads.   For eccentric loads on slab bridges, the variations of longitudinal moments and shears near their respective peaks are not as sharp as for loads placed well away from the free edges. Because of this phenomenon, the peak intensities of longitudinal moments and shears due to eccentric loads are relatively insensitive to minor changes in the spacings of longitudinal beams of the idealization and can be determined fairly accurately by the customary method

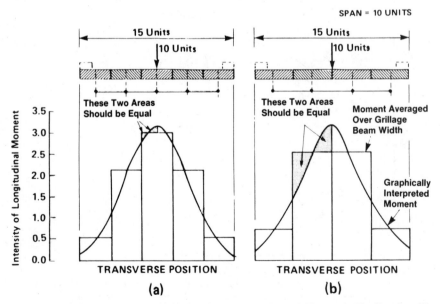

Figure 7.11 Proposed method of graphical interpretation. (*a*) Five-strip idealization. (*b*) Four-strip idealization.

mentioned earlier. However, there still remains the problem of obtaining integrated responses over a finite width of the slab. This problem can be solved conveniently by the following analytical procedure, which is adopted from Ref. 4.

The intensity of the longitudinal response (moment or shear) at a distance y from the free edge of the slab close to the loads is designated as A_y. Values of A_y corresponding to an eccentric loading are plotted in Fig. 7.12b across a transverse section of the bridge. The area under this curve over the width S represented by a longitudinal beam is

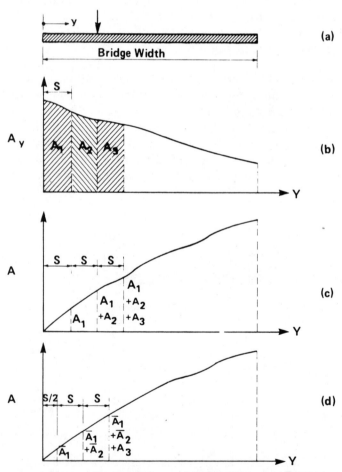

Figure 7.12 Transverse distribution of longitudinal responses due to eccentric loads. (a) Cross section. (b) Plot of intensities. (c) Plot of cumulative response for beams of the same rigidities. (d) Plot of cumulative response corresponding to outer beams of half of the rigidities of those of inner beams.

equal to the beam response. As shown in Fig. 7.12b, the beam responses are designated as A_1, A_2, A_3, etc. The cumulative response to the reference free edge is denoted as A.

The case of the idealization by equally spaced longitudinal beams of equal rigidities, as shown in the example of Fig. 7.2a, is first considered. As shown in Fig. 7.12c, the value of A at y equal to the beam spacing S is A_1, that at y equal to $2S$ is $(A_1 + A_2)$, and so on.

Analysis of a fairly large number of grillage idealizations of slab bridges have confirmed that the cumulative response curve for the case under consideration is well represented by the following equation:

$$A = C_1(y/S) + C_2(y/S)^2 + C_3(y/S)^3 \qquad (7.11)$$

where C_1 C_2, and C_3 are constants which depend upon the shape of the A curve; their values can be readily obtained if the value of A is known at three locations. Substituting A_1, $(A_1 + A_2)$, and $(A_1 + A_2 + A_3)$ at y equal to S, $2S$, and $3S$ respectively in Eq. (7.11) gives the following three expressions for the constants:

$$C_1 = (11A_1 - 7A_2 + 2A_3)/6$$

$$C_2 = (-6A_1 + 9A_2 - 3A_3)/6 \qquad (7.12)$$

$$C_3 = (A_1 - 2A_2 + A_3)/6$$

When a slab is idealized by equally spaced longitudinal beams, the outer ones of which represent a slab width of $S/2$, as for the example shown in Fig. 7.2b, the values of the beam responses are designated as $\overline{A}_1 + \overline{A}_2 + \overline{A}_3$, etc. In this case, as shown in Fig. 7.12d, the values of A at y equal to $0.5S$, $1.5S$, and $2.5S$ are $\overline{A}_1$, $(\overline{A}_1 + \overline{A}_2)$ and $(\overline{A}_1 + \overline{A}_2 + \overline{A}_3)$ respectively. It can be shown readily that by substituting these values of A in Eq. (7.11) the following expressions are obtained for the constants:

$$C_1 = (184\overline{A}_1 - 41\overline{A}_2 + 9\overline{A}_3)/60$$

$$C_2 = (-12\overline{A}_1 + 8\overline{A}_2 - 2\overline{A}_3)/5 \qquad (7.13)$$

$$C_3 = (8\overline{A}_1 - 7\overline{A}_2 + 3\overline{A}_3)/15$$

Since A is the response intensity A_y integrated along y, A_y can be obtained by differentiating A with respect to y, i.e.,

$$A_y = \frac{dA}{dy} \qquad (7.14)$$

or from Eq. (7.11),

$$A_y = \frac{C_1}{S} + \frac{2C_2}{S^2} y + \frac{3C_3}{S^3} y^2 \qquad (7.15)$$

The value of response A_y integrated over values of y from w_1 and w_2 is denoted as $A_{w1,w2}$; this quantity can be obtained directly from Eq. (7.11) as

$$A_{w1,w2} = C_1 \left(\frac{w_2 - w_1}{S}\right) + C_2 \left(\frac{w_2 - w_1}{S}\right)^2 + C_3 \left(\frac{w_2 - w_1}{S}\right)^3 \qquad (7.16)$$

It is noted that Eqs. (7.15) and (7.16) should be used only in the vicinity of the three outer beams whose responses are employed in the determination of the constants.

The application of the proposed analytical procedure is illustrated in the following with the help of a specific problem of a square slab which is analyzed by the semicontinuum method of Chap. 4 by using four different idealizations. Details of the slab, its four idealizations, and the applied loading are given in Fig. 7.13. The units are omitted from this figure to maintain generality. It is required to calculate the longitudinal shear at section xx shown in Fig. 7.13, sustained by an outer eight-unit-wide strip of the slab.

The values of shears at section xx in the three outer beams, i.e., A_1, A_2, and A_3 respectively, as obtained by the semicontinuum method, are listed in Table 7.1 for the four idealizations. The table also lists the values of constants C_1, C_2, and C_3 as obtained by Eq. (7.12). By

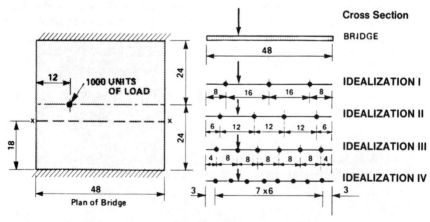

Figure 7.13 Details of a bridge and its idealization by the semicontinuum method.

TABLE 7.1   Values of Beam Shears and Constants at Section *xx* Shown in Fig. 7.13

Idealization	A_1	A_2	A_3	C_1	C_2	C_3
I	355.8	100.7	43.4	549.3	−226.5	33.0
II	302.7	115.6	50.8	437.0	−154.7	20.4
III	207.9	140.9	66.4	238.9	−29.8	−1.2
IV	156.8	128.0	81.8	165.4	−5.7	−2.9

using these values of constants and substituting 0.0 and 8.0 for w_1 and w_2 respectively in Eq. (7.16), the values of integrated longitudinal shear for idealizations I, II, III, and IV are found to be 1521, 1442, 1449, and 1438 units respectively. It is interesting to note that, with the help of the proposed procedure, even the coarse idealization I gave the value of the required response to be within about 6 percent of that by the fine idealization IV. This observation confirms that fairly reliable results can be obtained even with an idealization incorporating relatively few longitudinal beams.

Analytical procedure for loads in the central region. As discussed earlier, the transverse distribution of the intensities of longitudinal responses, i.e., M_x and V_x, is peakier for loads well away from the free edges of the slab than it is for loads near the free edges. Because of this phenomenon, the peaks of M_x and V_x due to loads in the central region become very sensitive to the assumed shape of the transverse distribution curve, such as that represented by Eq. (7.11). When there is a single concentrated load in the middle region of the slab, the transverse distribution curve in the vicinity of the load is at its peakiest. In such a case it is advisable to use the graphical method given earlier in the section to obtain the peak intensity. However, the following analytical method can be used to determine the peak intensity when there is more than one concentrated load in the middle region.

The case first considered for loads well away from the free edges of the slab is one in which a load sits directly on the beam having the maximum response, so that it can be realistically assumed that the peak intensity of the response lies directly under this beam. The cross sections of the bridge and the idealized semicontinuum are shown in Fig. 7.14a and b respectively. The responses of beams in the vicinity of the beam under consideration are shown in histogram form in Fig. 7.14c. As shown in this figure, the beam responses are designated as A_1, A_2, $\overline{A}_2$, A_3, and $\overline{A}_3$.

As shown in Fig. 7.14d, the cumulative responses A and $\overline{A}$ are measured from either side of the beam having the maximum response. The distance to the right side of this reference beam is defined as y, and that to the left as $\bar{y}$. As for eccentric loads, it is assumed that the shape of cumulative response curves can be defined by the following equations, which involve the responses of three beams in each case.

$$A = C_1(y/S) + C_2(y/S)^2 + C_3(y/S)^3 \qquad (7.17)$$

$$\bar{A} = \bar{C}_1(\bar{y}/S) + \bar{C}_2(\bar{y}/S)^2 + \bar{C}_3(\bar{y}/S)^3 \qquad (7.18)$$

The intensity of the longitudinal response is equal to dA/dy or $d\bar{A}/d\bar{y}$, which at $y = 0$ and $\bar{y} = 0$ is equal to C_1/S. By solving five simultaneous equations it can be shown that for the case shown in Fig. 7.14 the value of the maximum intensity of the longitudinal response, designated as A_{max}, is given by

$$A_{max} = \frac{1}{30S}\left[46A_1 - \frac{41}{4}(A_2 + \bar{A}_2) + \frac{9}{4}(A_3 + \bar{A}_3)\right] \qquad (7.19)$$

The second case for loads in the middle region is one in which the peak intensity of a longitudinal response is slightly offset from the beam location. This can happen if a concentrated load, instead of sitting

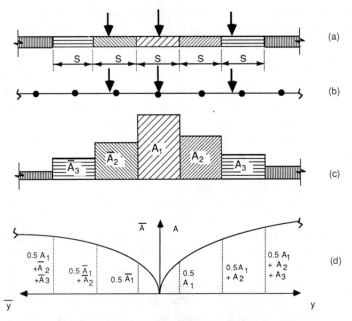

Figure 7.14 Transverse distribution of a longitudinal response. (*a*) Cross section of the actual slab. (*b*) Cross section of the semicontinuum. (*c*) Plot of average intensities. (*d*) Plot of cumulative intensity.

directly above the beam having the maximum response, is offset from it. It is reasonable to assume that the peak intensity of a response lies directly below the concentrated load closest to the beam having the maximum response. As shown in Fig. 7.15a, the distance of this concentrated load is defined as gS from the left-hand edge of the strip represented by the beam having the maximum response.

Other notation is kept the same as for the previous case, except that the common origin for y and $\bar{y}$ is moved under the concentrated load, as shown in Fig. 7.15d. As for the previous case, the cumulative responses A and $\bar{A}$ are assumed to be given by Eqs. (7.17) and (7.18) respectively. The peak intensity $A_{\max}$ of the response can again be shown to be equal to C_1/S. By solving the set of five simultaneous equations for the cumulative response, $A_{\max}$ is given by the following expression:

$$A_{\max} = \frac{A_1 - k_g A_2 - k_{1-g}\bar{A}_2 - l_g A_3 - l_{1-g}\bar{A}_3}{S(1 - m_g - m_{1-g})} \qquad (7.20)$$

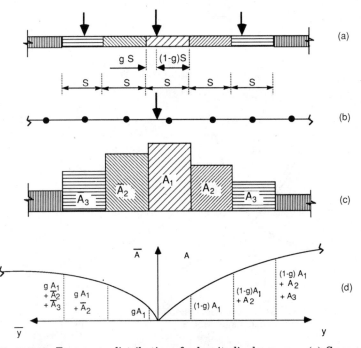

Figure 7.15 Transverse distribution of a longitudinal response. (a) Cross section of the actual slab. (b) Cross section of the semicontinuum. (c) Plot of average intensities. (d) Plot of cumulative intensity.

where

$$k_g = \frac{g^2(7 + 6g + g^2)}{4 + 12g + 6g^2}$$

$$k_{1-g} = \frac{(1 - g)^2[7 + 6(1 - g) + (1 - g)^2]}{4 + 12(1 - g) + 6(1 - g)^2}$$

$$l_g = -\frac{g^2(1 + 2g + g^2)}{4 + 12g + 6g^2}$$

(7.21)

$$l_{1-g} = -\frac{(1 - g)^2[1 + 2(1 - g) + (1 - g)^2]}{4 + 12(1 - g) + 6(1 - g)^2}$$

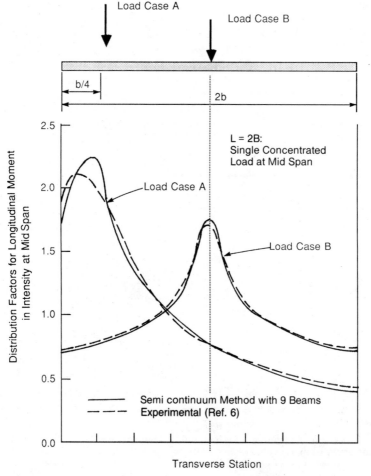

Figure 7.16 Comparison of distribution factors for longitudinal moments.

$$m_g = \frac{g^2(6 + 4g)}{4 + 12g + 6g^2}$$

$$m_{1-g} = \frac{(1 - g)^2[6 + 4(1 - g)]}{4 + 12(1 - g) + 6(1 - g)^2}$$

7.3   Verification of Proposed Method

The validity of the proposed semicontinuum method for the analysis of slab bridges was established by comparing its results with those given by already tested isotropic-slab methods and also by those obtained from model tests. One such comparison is given in Fig. 7.16 for distribution coefficients for longitudinal moments in a square concrete slab. In this figure, the distribution coefficients for midspan longitudinal moments due to two separate concentrated loads as given by the semicontinuum method are compared with experimentally obtained distribution coefficients reported in Ref. 6. The semicontinuum analysis was carried out by idealizing the square slab into nine equally spaced longitudinal beams with each outer beam having one-half of the rigidity of the inner ones and by using only the first three harmonics. The graphical methods were used for obtaining the peak intensities of the moments. As recommended in Sec. 7.2.1, no adjustment was made in the longitudinal moments for the effects of Poisson's ratio.

As can be seen in Fig. 7.16, there is excellent correlation between the semicontinuum and experimental results, thus confirming the validity of the former.

REFERENCES

1. Cusens, A. R., and Pama, R. P.: *Bridge Deck Analysis,* Wiley, London, 1975.
2. Elliott, G., and Clark, L. A.: Circular voided slab stiffnesses, *Journal of the Structural Division, ASCE,* 108(ST11), 1982, pp. 2379–2393.
3. Jaeger, L. G., and Bakht, B.: The grillage analogy in bridge analysis, *Canadian Journal of Civil Engineering,* 9(2), 1982, pp. 224–235.
4. Jaeger, L. G., and Bakht, B.: On the analysis of slab bridges by grillage analogy, *Proceedings, Annual Meeting of the Canadian Society for Civil Engineering, Montreal,* May 1987, pp. 51–63.
5. *Ontario Highway Bridge Design Code,* Ministry of Transportation of Ontario, Downsview, Ontario, 1983.
6. Rowe, R. E.: *Load Distribution in Bridge Slabs (with Special Reference to Transverse Bending Moments Determined from Tests on Three Pre-stressed Concrete Slabs),* Cement and Concrete Association Technical Report TRA/199, London, 1956.

Rigid Frame Bridges

Figure 8.1 shows the longitudinal and transverse sections of a typical rigid frame bridge. In such bridges the girders are rigidly joined to supporting legs, which are usually vertical or nearly vertical in orientation. The girders are integral with a concrete deck slab. The composition of the bridge can be thought of as being similar to a number of portal frames connected with a concrete deck slab. Usually, the supporting legs are restrained from movement by the earth fill behind them; this restraint means that the overall sway of the portal frames is so small as to be negligible. The restraint also provides a high degree of rotational fixity at the girder ends.

It is shown in this chapter that rigid frame bridges with prismatic and equally spaced girders can also be analyzed for load distribution by making minor adjustments to the semicontinuum method of Chap. 4. For such analysis, the girders are assumed to be rotationally fixed at their ends.

8.1 Basis of Analysis

To analyze a rigid frame bridge, the selected idealization of the bridge is once again the semicontinuum in which the longitudinal flexural and torsional stiffnesses are assumed to be concentrated into a number of discrete line elements, while the transverse stiffnesses are taken to be uniformly distributed among an infinite number of elementary transverse beams which thereby constitute a transverse medium. The discrete longitudinal line elements are taken to be at the positions of the actual girders of the bridge, thus giving a close approximation of the real bridge by the analytical model.

The power of the semicontinuum model in reducing the number of unknowns derives from the use of characterizing load shapes having the basic property that an applied load of that shape will provide a deflection similar to itself, as discussed in Chaps. 3 and 4 and in Refs. 2 and 3. In the case of a longitudinal girder which is simply supported at its ends, the characterizing shapes are sine waves, and a full treatment of the transverse distribution analysis of simply supported bridges has been given in Chaps. 3 and 4.

When the girders of a bridge are taken to be fully fixed against rotation at their ends, the set of characterizing shapes is slightly more complex than the simple sine waves, but the same general principles of load distribution apply. The methods for obtaining these shapes are given in App. I. Figure 8.2a shows a fixed-ended beam carrying a distributed load in the shape of the first of the characterizing functions. The deflections and bending moments arising from this loading are shown in Fig. 8.2b and c respectively. It can be seen that the beam deflections are of the same shape as the applied loading. The shape of the deflection profile is denoted by $X_1(x)$ and the shape of the bending-moment profile by $\Phi_1(x)$.

If a line loading in the shape of the first characterizing function $X_1(x)$ is applied to one of the girders of a rigid frame bridge with full rotational fixity at both ends of each girder, then it is readily demonstrated that, provided that a simplifying assumption is made about torsional

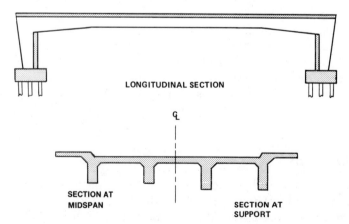

Figure 8.1 Details of a typical rigid frame bridge.

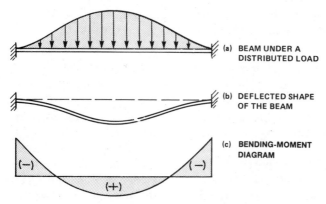

(a) BEAM UNDER A
    DISTRIBUTED LOAD

(b) DEFLECTED SHAPE
    OF THE BEAM

(c) BENDING-MOMENT
    DIAGRAM

Figure 8.2   Fixed-ended beam under a load distributed in the shape of a characterizing function.

behavior, the loads that are distributed to the other girders of the bridge are of the same shape as the applied line loading, and the proportions in which the externally applied load is shared by the various girders are readily calculated. For a rigid frame bridge having similar and equally spaced girders, the distribution coefficients depend upon the dimensionless parameters η, λ, and μ, which in the case of rotationally restrained ends are given in the following for loading due to the first harmonic:

$$\eta = \frac{12}{(1.5\pi)^4} \left(\frac{L}{S}\right)^3 \frac{LD_y}{EI}$$

$$\lambda = \frac{1}{(1.5\pi)^2} \left(\frac{L}{S}\right)^2 \frac{SD_{yx}}{EI} \tag{8.1}$$

$$\mu = \frac{1}{(1.5\pi)^2} \left(\frac{L}{S}\right)^2 \frac{GJ}{EI}$$

where L and S are span and girder spacing respectively, EI and GJ are the flexural and torsional rigidities of a girder respectively, and D_y and D_{yx} are transverse flexural and torsional rigidities per unit length of the deck slab respectively. The coefficients corresponding to loading by harmonic n are obtained by replacing π in the above equations by $n\pi$.

It may be noted that the definitions of η, λ, and μ given in Eq. (8.1) are identical with those for simply supported ends given in Eq. (4.36), except that the number π has been replaced by 1.5π throughout. This change reflects the difference in end conditions. The replacement of π by 1.5π means that the numerical values of η, λ, and μ are different from what they would be with simply supported ends, so that the distribution coefficients are also different.

A simplifying treatment of torsional effects. Excluding bridges with girders of hollow sections from consideration, the smallest values of torsional parameters λ and μ are for timber bridges and the largest for solid-slab bridges. It has been shown in Ref. 1 that for torsional stiffnesses within this range the values of distribution coefficients depend more on the sum of the torsional parameters than on their individual values. This is particularly true for bridges with relatively small torsional stiffnesses such as slab-on-girder bridges with steel girders or solid concrete girders. To facilitate the analysis of rigid frame bridges, the torsional stiffness of the deck slab is assumed to be zero, and its value is added to the torsional stiffness of the longitudinal beams. By using this simplifying assumption, an external line load $PX_1(x)$ is applied to the bridge, where $X_1(x)$ is the load shape shown in Fig. 8.2a and P is an amplitude.

Figure 8.3 shows a unit strip of the transverse medium. The girders all deflect into the same shape $X_1(x)$, which should be compared with the shape $\sin(\pi x/L)$ for simply supported ends. Girder 1 undergoes deflection y_1, given by

$$y_1 = a_1 X_1(x) \tag{8.2}$$

and rotation ϕ_1, given by

$$\phi_1 = v_1 X_1(x) \tag{8.3}$$

and similarly for the other girders.

The loading on girder 1 to produce the deflection of Eq. (8.2) is

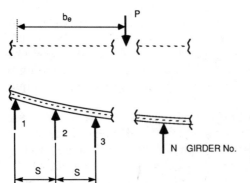

N   GIRDER No.   **Figure 8.3** Deflected shape of a strip of the transverse medium.

$ka_1 X_1(x)$, where

$$k = (1.5\pi)^4 \frac{EI}{L^4} \tag{8.4}$$

In Eq. (8.4) the presence of 1.5π in place of π in the corresponding formula for simply supported ends is again noted.

The torsional stiffness of the transverse medium is put to zero in the governing equations, i.e., $\lambda = 0$, and the torsional-stiffness parameter in the longitudinal direction is increased from μ to $(\lambda + \mu)$, which increased parameter is given the symbol β. Then for the simplified treatment of rigid frame bridges only two parameters, η and β, are used where, for first harmonic loading,

$$\eta = \frac{12}{(1.5\pi)^4} \left(\frac{L}{S}\right)^3 \frac{LD_y}{EI} \tag{8.5}$$

$$\beta = \frac{1}{(1.5\pi)^2} \left(\frac{L}{S}\right)^2 \frac{GJ + SD_{yx}}{EI}$$

Then analysis of the load case shown in Fig. 8.3 proceeds in exactly the same way as for simply supported ends. The shape factor $X_1(x)$ cancels through as a common factor, and the distribution coefficients are obtained by the same expressions as given in Chap. 10 for bridges with simply supported ends. The technique of combining λ and μ into β is also used in Chap. 10 to make possible the analysis of simply supported bridges by manual calculations. That chapter contains the expressions and charts for the various distribution coefficients in three-, four-, and five-girder bridges. The expressions are given in Eqs. (10.21) through (10.29) and the charts in Figs. 10.3, 10.4, and 10.5. These expressions and charts can also be used for rigid frame bridges provided that η and β are calculated from Eq. (8.5).

Free-bending-moment diagram. The "free" bending moment M_L is calculated from the magnitude and position of applied load from simple beam theory. This is the bending moment which the loaded girder would sustain if no transverse distribution to other girders took place. For the fixed-ended beam M_L is given by

$$M_L = M_L^{(1)}\Phi_1(x) + M_L^{(2)}\Phi_2(x) + \cdots \tag{8.6}$$

The procedure for obtaining $M_L^{(1)}$, $M_L^{(2)}$, etc., is given in App. I, where it is shown that $M_L^{(1)}$ is given by

$$M_L^{(1)} = \frac{0.9653}{L} \int_0^L M_L \Phi_1(x)\, dx \tag{8.7}$$

For evaluating the integral in Eq. (8.7) numerically, the values of $\Phi_1(x)$ for different values of x are listed in Table I.1.

The only difference in the analysis of fixed-ended and simply supported bridges lies in the calculation of the characterizing parameters and calculation of $M_L^{(1)}$, $M_L^{(2)}$, etc. The procedure for analyzing rigid frame bridges is illustrated in the following example by considering only the first harmonic and assuming that the loads due to second and higher harmonics are retained entirely by the externally loaded girders. The procedure thus becomes similar to that of the manual method for simply supported bridges given in Chap. 10.

8.2 Worked Example

Figure 8.4 shows the cross section of a four-girder rigid frame bridge which carries one line of wheels of the AASHTO MS 18 vehicle on an outer girder. The girders are prismatic; i.e., they have a uniform flexural rigidity along the spans. As shown in Figs. 8.4 and 8.5, the span of the bridge and girder spacing are 20.0 and 2.4 m respectively. The longitudinal positions of the wheels and their magnitudes are shown in Fig. 8.5. The unlikely condition of a single line of wheels on the bridge is, of course, taken for simplicity to illustrate the use of the method. The modulus of elasticity E and shear modulus G of both the girder and deck-slab material are assumed to be 20×10^6 and 9×10^6 kN/m².

The transverse rigidity D_y is contributed by only the deck slab and has the value $Et^3/12$, where E is the modulus of elasticity of the concrete of the deck slab and t is the slab thickness. For a slab thickness of 200 mm, the value of D_y is found to be 0.013×10^6 (kN·m²)/m.

The flexural rigidity EI of a line member representing a girder and the associated portion of the deck slab can be found by standard meth-

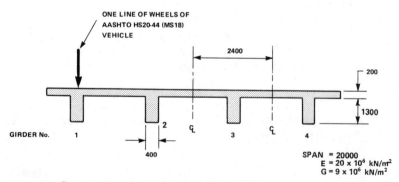

Figure 8.4 Cross section of a rigid frame bridge. All dimensions are in millimeters.

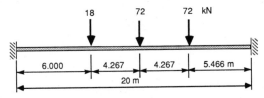

Figure 8.5 Longitudinal position of line of wheels.

ods (e.g., those given in Refs. 1 and 4) to be 4.304×10^6 kN·m². The sum of $S\,D_{yx}$ and the longitudinal torsional rigidity GJ of a girder and the associated portion of the deck slab is found to be equal to 0.270×10^6 kN·m². Thus from Eq. (8.5),

$$\eta = \frac{12}{(1.5\pi)^4} \left(\frac{20.0}{2.4}\right)^3 \frac{20.0 \times 0.013 \times 10^6}{4.304 \times 10^6} = 0.85$$

$$\beta = \frac{1}{(1.5\pi)^2} \left(\frac{20.0}{2.4}\right)^2 \frac{0.270 \times 10^6}{4.304 \times 10^6} = 0.20$$

The distribution coefficients for these values of η and β can be read directly from Fig. 10.4 for four-girder bridges. Alternatively, they may be calculated from Eqs. (10.23) to (10.25). The coefficients obtained from Fig. 10.4 for load on girder 1 are listed in Table 8.1.

The values of M_L are obtained from beam analysis for various values of x; these values of M_L are listed in Table 8.2 together with the corresponding values of $\Phi_1(x)$, which are taken from Table I.1.

By using the values of M_L and $\Phi_1(x)$ listed in Table 8.2, the value of $M_L^{(1)}$ is obtained as follows from Eq. (8.7) by employing numerical integration with the familiar Simpson's rule.

$$M_L^{(1)} = \frac{0.9653}{L} \times \frac{L}{10 \times 3} [(-306.1 \times -2.036)$$

$$+ 4(-182.4 \times -1.094) + 2(-58.7 \times -0.199) + \cdots$$

$$+ 4(-214.9 \times -1.094) + (-415.2 \times -2.036)]$$

$$= 181.7 \text{ kN·m}$$

TABLE 8.1 Distribution Coefficients Corresponding to $\eta = 0.85$, $\beta = 0.20$, for Load on Girder 1

Coefficient	$\rho_{1,1}$	$\rho_{2,1}$	$\rho_{3,1}$	$\rho_{4,1}$
Value	0.82	0.19	0.00	−0.01

TABLE 8.2 Values of M_L and $\Phi_1(x)$

x,m	0	2	4	6	8	10	12	14	16	18	20
M_L, kN·m	−306.1	−182.4	−58.7	65.1	152.8	240.5	203.5	147.2	−14.7	−214.9	−415.2
$\Phi_1(x)$	−2.036	−1.094	−0.199	0.553	1.059	1.237	1.059	0.553	−0.199	−1.094	−2.036

TABLE 8.3 Values of M_L and $M_L^{(1)}\Phi_1(x)$

x,m	0	2	4	6	8	10	12	14	16	18	20
M_L, kN·m	−306.1	−182.4	−58.7	65.1	152.8	240.5	203.5	147.2	−14.7	−214.9	−415.2
$M_L^{(1)}\Phi_1(x)$, kN·m	−370.0	−198.8	−36.1	100.5	192.5	224.8	192.5	100.5	−36.1	−198.8	−370.0

TABLE 8.4 Steps of Calculations in Obtaining Moments Retained by the Loaded Girder

Row no.	x	0	4	8	10	12	16	20
					m			
1	Free moment, M_L, kN·m	−306.1	−58.7	152.8	240.5	203.5	−14.7	−415.2
2	First harmonic component of free moment, $M_L^{(1)}\Phi_1(x)$, kN·m	−370.0	−36.1	192.5	224.8	192.5	−36.1	−370.0
3	Moment transferred to girder 2, $M_L^{(1)}\Phi_1(x)$, kN·m ($\rho_{2,1} = 0.19$)	−70.3	−6.9	36.6	42.7	36.6	−6.9	−70.3
4	Moment transferred to girder 3, $\rho_{3,1} \times M_L^{(1)}\Phi_1(x)$, kN·m ($\rho_{3,1} = 0.00$)	0.0	0.0	0.0	0.0	0.0	0.0	0.0
5	Moment transferred to girder 4, $\rho_{4,1} \times M_L^{(1)}\Phi_1(x)$, kN·m ($\rho_{4,1} = 0.01$)	3.7	0.4	−1.9	−2.2	−1.9	0.4	3.7
6	Sum of moments transferred to girders 2, 3, and 4, kN·m	−66.6	−6.5	34.7	40.5	34.3	−6.5	−66.6
7	Moments retained by girder 1 (= row 1–6), kN·m	−239.5	−52.2	118.1	200.0	169.2	−8.2	−348.8

Values of $M_L^{(1)}\Phi_1(x)$ are now found at the stated values of x, with the results shown in Table 8.3, which also lists the values of M_L for purposes of comparison. It can be seen that the function $M_L^{(1)}\Phi_1(x)$ is indeed the dominant component of M_L.

The various steps of calculating the moments retained by externally loaded girder 1 are given in Table 8.4. Row 1 contains the values of the free moment M_L at different values of x. Row 2 contains the values of the first harmonic component of loading $M_L^{(1)}\Phi_1(x)$, which are also listed in Table 8.3. Moments transferred to girders 2, 3, and 4 are obtained by multiplying the values of $M_L^{(1)}\Phi_1(x)$ by $\rho_{2,1}$, $\rho_{3,1}$, and $\rho_{4,1}$ respectively; the values of moments transferred to the three girders, obtained thus, are given in rows 3, 4, and 5 respectively. The sums of values in rows 3, 4, and 5 are equal to the moments that are passed away from the externally loaded girder; these sums are given in row 6. Now the net values of moments retained by the externally loaded girder are equal to the difference between the free moments, given in row 1, and the sum of moments passed away from the loaded girders, given in row 6. These net moments are given in row 7.

The above process of obtaining net moments in the externally loaded girder is illustrated in Fig. 8.6b, c, and d, the first of which shows the free-moment diagram and the second the diagram of moments transferred from the loaded girder to the other girders. The net moments in girder 1, corresponding to the values given in row 7 of Table 8.4, are plotted in Fig. 8.6d.

It is interesting and instructive to note that the ratio of the moment accepted by a girder to the free moment does not remain constant along the span. For example, in girder 1 this ratio is equal to 0.78 at the left-hand support, 0.83 at midspan, and 0.84 at the right-hand support.

Comparison with grillage analysis. The bridge of the worked example was also analyzed by the grillage analysis method by idealizing it as an assembly of four longitudinal and nine transverse beams. Consistently with the boundary conditions inherent in the semicontinuum method for rigid frame bridges, the ends of the longitudinal beams were held against rotation. It was found that the results of grillage analysis agreed very closely with the corresponding semicontinuum analysis results. Moments in the externally loaded girder by the two methods of analysis are compared in Fig. 8.6d. The close agreement of the two moment diagrams confirms the validity of the semicontinuum method given in this chapter for the analysis of rigid frame bridges with rotationally fixed ends.

Note of caution. It is emphasized that the method of analysis presented in this chapter can deal with only those rigid frame bridges in which the longitudinal flexural rigidities remain substantially constant along the span. Unlike the case for simply supported bridges, the semicontinuum method cannot be readily adapted for rigid frame bridges whose longitudinal girders have significantly varying longitudinal flexural rigidity by the kind of approach presented in Sec. 5.3 for girders of varying flexural rigidity. It is recommended that such rigid frame bridges be analyzed by the grillage analogy or similar more general method.

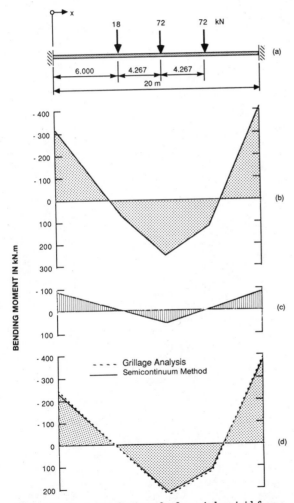

Figure 8.6 Load distribution of a four-girder rigid frame bridge. (a) Applied loading on girder 1. (b) Free-moment diagram. (c) Moments transferred to girders 2, 3, and 4. (d) Moments retained by girder 1.

REFERENCES

1. Bakht, B., and Jaeger, L. G.: *Bridge Analysis Simplified,* McGraw-Hill, New York, 1985.
2. Hendry, A. W., and Jaeger, L. G.: *The Analysis of Grid Frameworks and Related Structures,* Prentice-Hall, Englewood Cliffs, N.J., 1958.
3. Jaeger, L. G., and Bakht, B.: Bridge analysis by the semicontinuum method, *Canadian Journal of Civil Engineering,* 12(3), 1985, pp. 573–582.
4. *Ontario Highway Bridge Design Code,* Ministry of Transportation and Communications, Downsview, Ontario, 1983.

Computer Programs

Several methods of analysis presented in this book are incorporated into computer programs which are written in machine-independent standard FORTRAN and can be run either on a microcomputer or on a mainframe computer. Program manuals and user manuals for these programs are given in this chapter.

9.1 Program for Multispan Beams, CONBIM

The procedure described in Sec. 2.4 for harmonic analysis of multispan continuous beams is incorporated in a program called CONBIM. This program analyzes multispan continuous beams of uniform flexural rigidity subjected to a series of point loads. It should be noted, however, that while the procedure given in Sec. 2.4 permits the intermediate supports to be elastic and to have prescribed settlements, the program CONBIM can deal only with those intermediate supports which are fixed against deflection.

CONBIM calculates beam deflections, moments, and shears at required reference points; it automatically selects the number of harmonics so that the convergence of results is in accordance with the criterion given in Sec. 2.4. It is recalled that, according to this criterion, the beam responses are assumed to have converged fully if the values of intermediate-support reactions vary by less than 0.1 percent for three successive harmonics. The listings of CONBIM are given in App. V.

9.1.1 Program manual for CONBIM

Descriptions of subroutines. CONBIM consists of the following subroutines.

1. Subroutine AMATRX forms the $[A]$ matrix, which is defined by Eqs. (2.9) and (2.10).
2. Subroutine DVECT forms the $\{D\}$ vector, which is defined by Eqs. (2.9) and (2.10).
3. Subroutine DEFLEC calculates deflections according to the last equation of Eq. (2.3) due to a given point load at a specified reference point for a given number of harmonics.
4. Subroutine EQN solves Eq. (2.9) to obtain the $\{R\}$ vector of intermediate-support reactions.
5. Subroutine REACT calculates the reactions of the two extreme simple supports of the continuous beam after the intermediate-support reactions have been calculated.
6. Subroutine REFW adds the deflections in a simply supported beam due to the various point loads.
7. Subroutine SHRAMO calculates shears and moments at given reference points after the intermediate-support reactions have been calculated.
8. The MAIN subroutine reads the input data, coordinates the various subroutines, and prints deflections, moments, and shears at the various reference points.

Limits of the program. The size limitations on the problem that the program can handle are to the designer's choice. In the standard formulation they are as follows.

1. Maximum number of intermediate supports = 4.
2. Maximum number of point loads = 10.
3. Maximum number of reference points at which moments and shears are required = 20.
4. Maximum number of reference points at which deflections are required = 20.

The above limits can be extended readily through the following simple changes in the program.

1. The limit on the number of intermediate supports can be increased from 4 to N by increasing the dimensions of arrays XCOL, PIVOT, DD,

PD, D, B, IPIVOT, and COUNT from 4 to N; of array A from (4×4) to $(N \times N)$ and of array INDEX from (4×2) to $(N \times 2)$.

2. The limit on the number of point loads can be extended by increasing the dimensions of arrays XLOAD and PLOAD to the new number of point loads.

3. The maximum number of reference points at which moments and shears are required can be extended by changing the dimensions of arrays X, SHR, and MOM to the new number of reference points.

4. Similarly, the maximum number of reference points at which deflections are required can be extended by changing the dimensions of arrays XREFW and SUMW to the new number of reference points.

It is noted that the above changes to the array sizes must be made in all the subroutines where they occur.

9.1.2   User manual for CONBIM

Data input.   The data input, which is in free format, is required in the following sequence. All input may be in any set of consistent units.

1. Span length, EI of the beam, number of applied point loads, number of intermediate supports, number of reference points at which moments and shears are required, and number of reference points at which deflections are required. It is recalled that the distance between the two extreme simple supports of the beams is regarded here as the span length and that EI is the flexural rigidity of the beam.

2. Distances of all applied point loads from the left-hand simple support in ascending order.

3. Magnitudes of all applied point loads in the same order as their distances are given in Par. 2.

4. Distances of all intermediate supports from the left-hand simple support in ascending order.

5. Distances of all reference points at which moments and shears are required from the left-hand simple support in ascending order.

6. Distances of all reference points at which deflections are required from the left-hand simple support in ascending order.

Output.   The output of CONBIM consists of a partial echo print of the input data along with separate listings of moments and shears and of deflections at the stated reference stations. It also gives the number of harmonics required to achieve the stated convergence. It is noted that the very small values of moments and deflections that may be given at the two simple supports, which are fixed against vertical deflections, are the result of rounding-off errors.

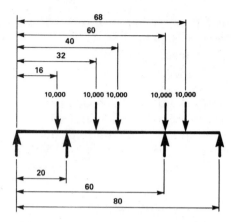

Figure 9.1 A three-span beam.

Example. A three-span beam, i.e., one with two intermediate supports, is analyzed by CONBIM. As shown in Fig. 9.1, the continuous beam is subjected to five point loads. The data input for the problem is as follows.

```
80   0.1E11    5     2     8   13
16       32    40    60    68
1E4      1E4   1E4   1E4   1E4
20       60
 0       16    20    32    40  60  68  80
 0       16    20    32    40  45  50  55  60  65  68  75  80
```

From the output shown in Fig. 9.2 it can be seen that a total of nine harmonics were required for the intermediate-support reactions to converge according to the previously stated criterion. The reactions of the first and second intermediate supports, counting from the left, are 24,475 and 27,273 units respectively, compared with true reactions of 24,470 and 27,270 units respectively. It can be seen that the values predicted by CONBIM are within 0.02 percent of the true values.

9.2 Program for Simply Supported Bridges, SECAN1

SECAN is an acronym derived from *semicontinuum analysis*, and SECAN1 is the name of the program which analyzes simply supported right bridges for lateral load distribution by the semicontinuum method. The program discussed in this section is based upon the theory presented in Chap. 4.

For a right simply supported bridge, which can be idealized as an assembly of one-dimensional longitudinal beams and a transverse medium and which is subjected to a series of concentrated loads, SECAN1

```
INPUT DATA
SPAN =  80.0          EI = 0.100E+11

COLUMN      DIST. FROM LEFT
  1              20.0
  2              60.0

LOAD        DIST. FROM LEFT
 10000.          16.0
 10000.          32.0
 10000.          40.0
 10000.          60.0
 10000.          68.0
```

****

9 HARMONICS REQUIRED FOR CONVERGENCE

```
COLUMN      REACTION
  1            24475.
  2            27273.
```

LEFT-END REACTION   RIGHT-END REACTION
```
        -2174.               427.
```

```
DIST. FROM LEFT      SHEAR           MOMENT
     0.0             -2174.             0.
    16.00            -2174.         -34791.
    20.00           -12174.         -83488.
    32.00            12301.          64119.
    40.00             2301.          82524.
    60.00            -7699.         -71464.
    68.00             9573.           5120.
    80.00             -427.             -2.

DIST. FROM LEFT      DEFLECTION
     0.0             0.0
    15.00            -.109E-03
    20.00            0.186E-07
    32.00            0.679E-03
    40.00            0.832E-03
    45.00            0.697E-03
    50.00            0.445E-03
    55.00            0.175E-03
    60.00            0.186E-07
    65.00            -.444E-04
    68.00            -.350E-04
    75.00            -.698E-05
    80.00            -.127E-09
```

Figure 9.2 Output from CONBIM.

provides the values of most commonly required responses, namely, longitudinal moments, longitudinal shears, and deflections. The listings of this program are given in App. VI.

9.2.1 Program manual for SECAN1

The various subroutines of SECAN1 are arranged according to the flowchart shown in Fig. 9.3.

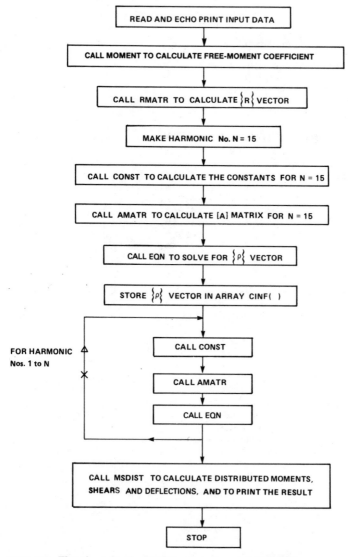

Figure 9.3 Flowchart for the MAIN subroutines of SECAN1.

Variables used in the program. Some of the variables used in the program are illustrated in Fig. 9.4. The following is a list of variables used for input and those used jointly by more than one subroutine.

ABM   Array containing transversely distributed bending moments. The first dimension refers to the girder number and the second to the reference section number.

ALFA   Array containing η_r for $r = 1$ to NG.

AMM   Free moment due to one longitudinal line of loads at stated distances from the left-hand simple support.

AS   Array containing transversely distributed longitudinal shears. The first dimension refers to the girder number and the second to the reference section number.

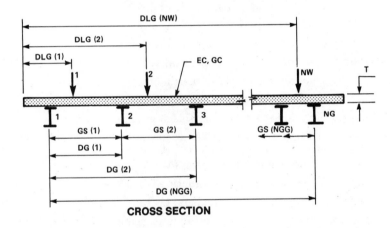

CROSS SECTION

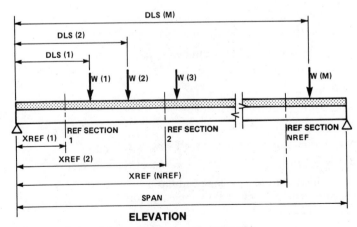

ELEVATION

Figure 9.4 Some of the variables used in SECAN1.

BM	Array containing coefficients for free moment due to one longitudinal line of wheels for each harmonic from 1 to N.
C	Array containing c_r for $r = 1$ to NG.
CINF	Array containing $\{\rho\}$ vector for harmonic 15.
DG	Array containing distances of girders 2 to NG from the left-hand outer girder.
DLG	Array containing distances of lines of loads from the left-hand outer girder, 1 to NW.
DLS	Array containing distances of loads in one longitudinal line from the left-hand simple support, 1 to M.
DY	D_y.
DYX	D_{yx}.
E	Modulus of elasticity of girder material.
EC	Modulus of elasticity of deck-slab material.
G	Shear modulus of girder material.
GC	Shear modulus of deck-slab material.
GMI	Array containing moments of inertia of girders 1 to NG.
GS	Array containing lengths of panels 1 to NGG.
GTI	Array containing torsional inertias of girders 1 to NG.
HARC	Array containing $\{\rho\}$ vectors for various harmonic numbers. The first dimension corresponds to the harmonic number and the second to the element numbers of $\{\rho\}$.
I1	Harmonic number used in subroutine CONST.
KT	Array containing k_r (real numbers) for $r = 1$ to NG.
KZ	Control number used for printing intermediate results; $= 1$ if intermediate results are not required; otherwise $= 2$.
LAMBDA	Array containing λ_r for $r = 1$ to NG.
M	Number of loads in one longitudinal line.
MLC	Array containing m_r (real numbers) for $r = 1$ to NG.
MU	Array containing μ_r (real numbers) for $r = 1$ to NG.
N	Number of harmonics.
NG	Number of girders.
NGG	Number of panels ($=$ NG $- 1$).
NREF	Number of reference sections.
NW	Number of longitudinal lines of loads.
RM	Array containing the ($2 \times$ NG) terms of the $\{R\}$ vector.
SHR	Free shear due to one longitudinal line of loads at a stated distance from the left-hand simple support.
SPAN	Span of the bridge.
T	Deck-slab thickness.

W   Array containing load magnitudes in a longitudinal line, 1 to M.

XREF   Array containing distances of reference sections from the left-hand simple support, 1 to NREF.

XX   Array containing the $\{\rho\}$ vector for a given harmonic number.

Description of subroutines.

1. MAIN *subroutine*.   This subroutine reads, echo-prints the input, and organizes the various subroutines according to the flowchart given in Fig. 9.3.

2. *Subroutine* AMATR.   This subroutine calculates all terms of the [A] matrix for each harmonic number according to the instructions of App. III. The various terms are stored in array AM (–,–). The dimensions of the matrix are (2NG × 2NG), NG being the number of girders.

3. *Subroutine* CONST.   This subroutine calculates the various constants, namely, k_r, m_r, c_r, λ_r, η_r, and μ_r, which are used in the construction of the [A] matrix. For each harmonic, r varies from 1 to the number of girders for each of the above constants.

4. *Subroutine* DEFLEC.   This subroutine calculates free deflections in a longitudinal beam due to applied point loads at a specified reference point.

5. *Subroutine* EQN.   This subroutine solves simultaneous equations:

$$[A]\{\rho\} = \{R\}$$

where the [A] matrix, having dimensions (2NG × 2NG), is stored in array AM (–,–); the $\{R\}$ vector, having 2NG elements, is stored in RM (–); and the solution vector $\{\rho\}$, which also has 2NG elements, is stored in array XX (–). The subroutine also stores the $\{\rho\}$ vectors in array HARC (–,–), the first dimension of which refers to the harmonic number and the second to the various elements of the vector.

6. *Subroutine* FINDEF.   This subroutine calculates deflections of longitudinal beams after taking account of the transverse distribution of loads.

7. *Subroutine* MOMENT.   This subroutine calculates coefficients for free moment due to one longitudinal line of loads. The coefficients are stored in array BM (–) separately for each harmonic number. Thus, for a given line of loads the free bending moment ML is given by

$$ML = BM\,(1)\,\sin\frac{\pi x}{L} + BM\,(2)\,\sin\frac{2\pi x}{L} + \cdots \qquad (9.1)$$

8. *Subroutine* MOMSER.   This subroutine calculates free moments and free shears due to one longitudinal line of loads in a beam at a

specified distance from the left-hand simple support. The moments and shears are stored in AMM and SHR respectively.

9. *Subroutine* MSDIST. This subroutine calculates the transversely distributed moments and shears at specified reference points due to a stated number of harmonics according to Eqs. (4.44) and (4.46) respectively. It calls subroutine MOMSER to obtain free moments and shears. Distributed moments and shears are stored in arrays ABM (–,–) and AS (–,–), where in both cases the first dimension refers to the girder number and the second to the reference section number. This subroutine also prints the transversely distributed moments and shears in all girders at the specified reference sections.

10. *Subroutine* RMATR. This subroutine calculates the 2NG terms of the $\{R\}$ vector for all lines of loads, where NG is the number of girders. The coefficients are stored in array RM (–).

Limits of the program.   The size limitations of the problem that can be analyzed by SECAN1 are as follows.

1. Maximum number of harmonics = 5.
2. Maximum number of girders = 10.
3. Maximum number of loads in one longitudinal line = 7.
4. Maximum number of longitudinal lines of loads = 10.
5. Maximum number of transverse reference sections = 10.

The above limits can be changed by altering the sizes of the various arrays which are identified in Tables 9.1 to 9.10 for the various subroutines. For example, if it is required to extend the limit for the maximum number of harmonics to 15, then the arrays in the various subroutines should be changed as shown below.

Subroutine	Revised arrays
MAIN	BM (15), HARC (15,20)
EQN	HARC (15,20)
FINDEF	BM (15), HARC (15,20)
MOMENT	BM (15)
MSDIST	BM (15), HARC (15,20)

The main routine also checks to see if the limits of the problem to be analyzed are exceeded; if they are, then an error message is printed to that effect and the execution stops. When the limits of the program are extended, then the statements which check these limits and which immediately follow the input statements should also be suitably changed.

TABLE 9.1 Dimensions of Arrays Used in MAIN Subroutine

Variables (showing dimensions currently used in program)	Alter dimensions shown below to change maximum number of				
	Harmonics from 5 to x_1	Girders from 10 to x_2	Loads in one longitudinal line of wheels from 7 to x_3	Longitudinal lines of loads from 10 to x_4	Reference sections from 10 to x_5
ABM (10,10)		(x_2, x_5)			(x_2, x_5)
ALFA (10)		(x_2)			
AM (20,20)		$(2x_2, 2x_2)$			
AS (10,10)		(x_2, x_5)			(x_2, x_5)
BM (5)	(x_1)				
C (10)		(x_2)			
CINF (20)		$(2x_2)$			
DG (10)		(x_2)			
DLG (10)				(x_4)	
DLS (7)			(x_3)		
GMI (10)		(x_2)			
GS (10)		(x_2)			
GTI (10)		(x_2)			
HARC (5,20)	$(x_1, 2x_2)$	$(x_1, 2x_2)$			
KT (10)		(x_2)			
LAMBDA (10)		(x_2)			
MLC (10)		(x_2)			
MU (10)		(x_2)			
NUMB (10)				(x_5)	
RM (20)		$(2x_2)$			
W (7)			(x_3)		
XREF (10)					(x_5)
XX (20)		$(2x_2)$			

TABLE 9.2 Dimensions of Arrays Used in Subroutine AMATR

Variables (showing dimensions currently used in program)	Alter dimensions shown below to change maximum number of girders from 10 to x_2
ALFA (10)	(x_2)
AM (20,20)	$(2x_2, 2x_2)$
C (10)	(x_2)
DG (10)	(x_2)
GS (10)	(x_2)
KT (10)	(x_2)
LAMDA (10)	(x_2)
MLC (10)	(x_2)
MU (10)	(x_2)

TABLE 9.3 Dimensions of Arrays Used in Subroutine CONST

Variables (showing dimensions currently used in program)	Alter dimensions shown below to change maximum number of girders from 10 to x_2
ALFA (10)	(x_2)
C (10)	(x_2)
GMI (10)	(x_2)
GS (10)	(x_2)
GTI (10)	(x_2)
KT (10)	(x_2)
LAMDA (10)	(x_2)
MLC (10)	(x_2)
MU (10)	(x_2)

TABLE 9.4 Dimensions of Arrays Used in Subroutine DEFLEC

Variables (showing dimensions currently used in program)	Alter dimensions shown below to change maximum number of	
	Girders from 10 to x_2	Loads in one longitudinal line of wheels from 7 to x_3
DLS (7)		(x_3)
GMI (10)	(x_2)	
W (7)		(x_3)

TABLE 9.5 Dimensions of Arrays Used in Subroutine EQN

Variables (showing dimensions currently used in program)	Alter dimensions shown below to change maximum number of	
	Harmonics from 5 to x_1	Girders from 10 to x_2
AM (20,20)		$(2x_2, 2x_2)$
BB (20,21)		$(2x_2, 2x_2 + 1)$
FF (20,21)		$(2x_2, 2x_2 + 1)$
HARC (5,20)	$(x_1, 2x_2)$	$(x_1, 2x_2)$
RM (20)		$(2x_2)$
SS (20,21)		$(2x_2, 2x_2 + 1)$
TT (20)		$(2x_2)$
XX (20)		$(2x_2)$

TABLE 9.6 Dimensions of Arrays Used in Subroutine FINDEF

Variables (showing dimensions currently used in program)	Alter dimensions shown below to change maximum number of			
	Harmonics from 5 to x_1	Girders from 10 to x_2	Loads in one longitudinal line of wheels from 7 to x_3	Reference sections from 10 to x_5
BM (5)	(x_1)			
CINF (20)		$(2x_2)$		
DLS (7)			(x_3)	
GMI (10)		(x_2)		
HARC (5,20)	$(x_1, 2x_2)$			
NUM (10)				(x_5)
XREF (10)				(x_5)
WF (10,10)		(x_2, x_5)		(x_2, x_5)

TABLE 9.7 Dimensions of Arrays Used in Subroutine MOMENT

Variables (showing dimensions currently used in program)	Alter dimensions shown below to change maximum number of	
	Harmonics from 5 to x_1	Loads in one longitudinal line from 7 to x_3
BM (5)	(x_1)	
DLS (7)		(x_3)
W (7)		(x_3)

TABLE 9.8 Dimensions of Arrays Used in Subroutine MOMSER

Variables (showing dimensions currently used in program)	Alter dimensions shown below to change maximum number of loads in one longitudinal line from 7 to x_3
DLS (7)	(x_3)
W (7)	(x_3)

TABLE 9.9 Dimensions of Arrays Used in Subroutine MSDIST

Variables (showing dimensions currently used in program)	Harmonics from 5 to x_1	Girders from 10 to x_2	Loads in one longitudinal line of wheels from 7 to x_3	Reference sections from 10 to x_5
		Alter dimensions shown below to change maximum number of		
ABM (10,10)		(x_2, x_5)		(x_2, x_5)
AMM2 (10)				(x_5)
AS (10,10)		(x_2, x_5)		(x_2, x_5)
BM (5)	(x_1)			
CINF (20)		$(2x_2)$		
DLS (7)			(x_3)	
HARC (5,20)	$(x_1, 2x_2)$	$(x_1, 2x_2)$		
NUM (10)				(x_5)
SHR2 (10)				(x_5)
W (7)			(x_3)	
XREF (10)				(x_5)

9.2.2 User manual for SECAN1

Data input. The input data for SECAN1, which can be in any set of compatible units, are in free format. The sequence of required input data is shown in the following along with an example for each set of entries, which corresponds to the bridge and loading shown in Fig. 9.5. It may be noted that, unlike the units used in the examples of the rest of the book, the units for the example of Fig. 9.5 are U.S. Customary System units. These are used to demonstrate that SECAN1 and other programs given in the book can deal with any compatible set of units.

1. Control number (= 2 if intermediate results are required for diagnostic purpose; otherwise = 1) and title in not more than 50 spaces. Example:

1 EXAMPLE OF FIG. 9.5

TABLE 9.10 Dimensions of Arrays Used in Subroutine RMATR

Variables (showing dimensions currently used in program)	Girders from 10 to x_2	Longitudinal lines of loads from 10 to x_4
	Alter dimensions shown below to change maximum number of	
DG (10)	(x_2)	
DLG (10)		(x_4)
GS (10)	(x_2)	
RM (20)	(x_2)	

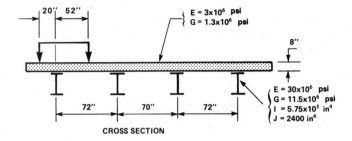

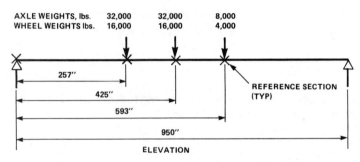

Figure 9.5 Details of a bridge and loading.

2. Number of harmonics with a maximum of 5, number of girders with a maximum of 10, span length, modulus of elasticity of girder material, and shear modulus of girder material. Example:

3	4	950	3E7	11.5E6

3. Girder spacings, starting from the left. Example:

72	70	72

4. Moment of inertia of all girders, starting from the outer left-hand girder. Example:

5.75E5	5.75E5	5.75E5	5.75E5

5. Torsional inertias of all girders, starting from the outer left-hand girder. Example:

2400	2400	2400	2400

6. Slab thickness, modulus of elasticity of slab material, and shear modulus of slab material. Example:

| 8 | 3E6 | 1.3E6 |

7. Number of concentrated loads in one longitudinal line with a maximum of 7. Example:

| 3 |

8. Magnitudes of concentrated loads in one longitudinal line, starting from the load closest to the left-hand simple support. Example:

| 1600 | 1600 | 4000 |

9. Distance of concentrated loads in one longitudinal line from the left-hand simple support. Example:

| 257 | 425 | 593 |

10. Number of longitudinal lines of concentrated loads with a maximum of 10. Example:

| 2 |

11. Transverse distances of longitudinal lines of concentrated loads from the outer left-hand girder. Example:

| −20 | 52 |

12. Number of transverse reference sections with a maximum of 10. Example:

| 4 |

13. Distances of reference sections from the left-hand abutment. Example:

| 0 | 257 | 425 | 593 |

The data input sheet for SECAN1 is given in Fig. 9.6 together with the limits of the various variables.

1.

Control No.	Title max. 52 characters

Control No. = 2 for diagnostic printing, otherwise = 1

2.

No. of harmonics (max. 5)	No. of girders (max. 10)	Span length	E of girder material	G of girder material

3. Girder spacings, starting from left

4. Moments of inertia of girders, starting from left

5. Torsional inertia of girders, starting from left

6.

Slab thickness	E of slab material	G of slab material

7.

No. of loads in one long-line (max. 7)

8. Weights of loads in one longitudinal line starting from left

9. Distances of loads in one longitudinal line from the left abutment

10.

No. of lines of loads (max. 10)

11. Transverse distances of lines of loads from the outer left girder

12.

No. of ref. sections (max. 10)

13. Distances of reference sections from the left-hand abutment

Figure 9.6 Data input sheet for SECAN1.

9.3 Program for Bridges with Intermediate Supports, SECAN2

The program SECAN2 is an extension of SECAN1; it can analyze bridges with simply supported ends and randomly spaced intermediate supports. A schematic plan for such a bridge is shown in Fig. 5.2.

SECAN2, which is based upon the theory developed in Chap. 5, can indeed handle continuous-span bridges as well, since these bridges are a particular case of the general category. The listings of this program are given in App. VII.

9.3.1 Program manual for SECAN2

SECAN2 has the same general structure as SECAN1, with the following sequence of operations in which SECAN1 can be regarded as a subprogram.

1. Read and reproduce the input data.

2. Remove intermediate supports, and use SECAN1 and subroutine WDIST to calculate deflections, longitudinal moments, and longitudinal shears at intermediate-support locations due to the applied loading for the prescribed number of harmonics. Also, obtain deflections, moments, and shears at the prescribed reference points.

3. Again, for the bridge without intermediate supports, obtain deflections at all the intermediate-support locations due to a unit load at the location of each intermediate support in turn.

4. From the input data on prescribed settlements of intermediate supports and the deflections obtained in operation 2, form the $\{D\}$ vector by calling subroutine DEE. It is recalled that the vector $\{D\}$ is defined by Eqs. (5.4) and (5.5).

5. From the input data on flexibility of the intermediate supports and the deflections obtained in operation 3, form the $[A]$ maxtix by calling subroutine BEE. It is noted that the $[A]$ matrix, which is defined by Eqs. (5.4) and (5.5), is renamed BI in SECAN2 to distinguish it from the $[A]$ maxtrix used in SECAN1.

6. Call the subroutine SOLVE to solve Eq. (5.5) and give the vector of reactions of the intermediate supports, i.e., vector $\{R\}$.

7. By treating the intermediate-support reactions as applied loads use SECAN1 and WDIST to calculate deflections, longitudinal moments, and longitudinal shears at the prescribed reference points.

8. Call the subroutine MSD2 to add the responses obtained in operations 2 and 7 to give the final responses.

9. Call the subroutine FINDEF to calculate net deflections due to applied loads and intermediate-support reactions.

Variables used in SECAN2. The same set of variables which are used in SECAN1 and described in Sec. 9.2.1 are used in SECAN2. In addition, the following variables are used jointly by several subroutines.

BI (–,–)   Array containing the $[A]$ matrix defined by Eqs. (5.4) and (5.5).

BMI (–)	Array containing coefficients for free moment due to a unit load at the location of an intermediate support.
DELTA (–)	Array containing the prescribed flexibilities of intermediate supports.
DI (–)	Array containing the $\{D\}$ vector defined by Eqs. (5.4) and (5.5).
DLSI (–)	The first term of this array contains the distance of the unit load at the location of intermediate support from the left-hand abutment.
FF (–)	Array containing prescribed flexibilities of intermediate supports.
HARC1 (–,–)	Same as HARC (–,–) but for unit loads at intermediate-support locations.
KGIR (–)	Girder numbers under which the various intermediate supports are located.
NCOL	Number of intermediate supports.
WU (–,–)	Array containing deflections due to unit loads at the various intermediate-support locations, with the first subscript referring to the support number for which deflections are sought and the second referring to the support number at which the unit load is applied.
WI (–)	The first term of this array is set equal to 1.0 to represent the unit load at an intermediate-support location.
XCOL (–)	Distances of the various intermediate supports from the left-hand abutment.

Description of subroutines. SECAN2 comprises the following subroutines in addition to those which constitute SECAN1 and are described in Sec. 9.2.1.

1. *Subroutine* BEE. This subroutine assembles the [A] matrix defined by Eqs. (5.4) and (5.5) from the input data on the flexibility of intermediate supports and the matrix of deflections WU (–,–). To distinguish this [A] matrix from the [A] matrix defined by Eq. (4.37) and used in the main subroutine, it is renamed BI.

2. *Subroutine* DEE. This subroutine assembles the $\{D\}$ vector of deflections at the intermediate-support locations, where $\{D\}$ is defined by Eqs. (5.4) and (5.5).

3. *Subroutine* DEFLEC. This subroutine calculates free deflections in a longitudinal beam due to the applied loading. Unlike subroutine MOMSER, which calculates free moments and shears by using simple beam theory, this subroutine calculates deflections by using harmonic analysis of beams for the number of harmonics specified in the input data.

4. *Subroutine* FINDEF.   This subroutine calculates the net deflections at the specified reference points due to applied loads and intermediate-support reactions.

5. *Subroutine* MSD2.   This subroutine calculates longitudinal moments and longitudinal shears due to the intermediate-support reactions and then adds them to the corresponding values obtained from MSDIST.

6. *Subroutine* WDIST.   This subroutine performs the same function as MSDIST except that it calculates deflections of longitudinal beams instead of moments and shears.

Limits of the program.   The limits on the size of the problem that can be handled by SECAN2 are the same as those for SECAN1. In addition, the maximum number of intermediate columns is 10. Problem size limits can be extended by the same procedure as specified for SECAN1 in Sec. 9.2.1 and by changing the arrays shown in Table 9.11 in all the relevant subroutines. It is recalled that, as in SECAN1, the MAIN subroutine of SECAN2 checks the limits of the size of the problem. Statements checking the limits should accordingly be altered when the limits are extended.

9.3.2   User manual for SECAN2

Data input.   The data input sheet for SECAN2 is given in Fig. 9.7. It can be seen that the data input for SECAN2 is the same as that for SECAN1 except for an addition in record 2 and for four additional

TABLE   9.11   Dimensions of Arrays Used in SECAN2 in Addition to Those Used in SECAN1

Variables (showing current dimensions)	Alter dimensions shown below to change the maximum number of				
	Harmonics from 5 to x_1	Girders from 10 to x_2	Loads in one longitudinal line of wheels from 7 to x_3	Longitudinal lines of loads from 10 to x_4	Number of intermediate supports from 10 to x_6
ARE (10)					(x_6)
BI (10,11)					$(x_6, x_6 + 1)$
BMI (5)	(x_1)				
DELTA (10)					(x_6)
DI (10)					(x_6)
DLGI (10)				(x_4)	
DLSI (7)			(x_3)		
FF (10)					(x_6)
HARC1 (5,20)	$(x_1, 2x_2)$	$(x_1, 2x_2)$			
WB (10)					(x_6)
WU (10)					(x_6)
W1 (7)			(x_3)		
XCOL (10,10)					(x_6, x_6)

1. | Control No. | Title max. 52 characters | | Control No. = 2 for diagnostic printing, otherwise = 1

2. | No. of harmonics (max. 5) | No. of girders (max. 10) | Span length | E of girder material | G of girder material | No. of inter. supports (max. 10) |

3. Girder spacings, starting from left

4. Moments of inertia of girders, starting from left

5. Torsional inertia of girders, starting from left

6. | Slab thickness | E of slab material | G of slab material |

7. No. of loads in one long-line (max. 7)

8. Weights of loads in one longitudinal line starting from left

9. Distances of loads in one longitudinal line from the left abutment

10. No. of lines of loads (max. 10)

11. Transverse distances of lines of loads from the outer left girder

12. No. of ref. sections (max. 10)

13. Distances of reference sections from the left-hand abutment

14. Prescribed deflections at intermediate support

15. Flexibility of intermediate support (in the same order as No. 14 above)

16. Girder No. under which each intermediate support is located (in the same order as No. 14 above)

17. Distance of each intermediate support from the left-hand abutment (in the same order as No. 14 above)

Figure 9.7 Data input sheet for SECAN2.

records after record 13. The records which differ from those of SECAN1 are given in the following paragraphs. Instructions for other records are as given in Sec. 9.2.2 for SECAN1.

2. Number of harmonics (maximum of 5), number of girders (maximum of 10), span length, modulus of elasticity of girder material, shear modulus of girder material, and number of intermediate supports (maximum of 10).

14. Prescribed settlements of intermediate supports in ascending order of support identification numbers. The same number of entries is required (even if the numbers are zero) as the number of intermediate supports.

15. Flexibilities of intermediate supports (length per unit force) in the same order as the entries in record 14.

16. Girder numbers under which the intermediate supports are located. The entries should be in the same ascending order of support identification numbers as in record 14.

17. Distances of intermediate supports from the left-hand abutment in the same order as entries in records 14 through 16.

Output. SECAN2, besides printing the input data with appropriate labels, provides the reactions at the intermediate supports and deflections, longitudinal moments, and longitudinal shears at the specified reference sections. If the control number in record 1 is set equal to 2,

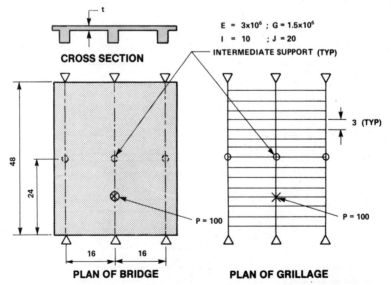

Figure 9.8 Details of bridge used in example. All dimensions are in compatible units.

1.

Control No.	Title max. 52 characters	Control No. = 2 for diagnostic printing, otherwise = 1
1	EXAMPLE	

2.

No. of harmonics (max. 5)	No. of girders (max. 10)	Span length	E of girder material	G of girder material	No. of inter. supports (max. 10)
3	3	48	3E6	1.5E6	3

3. Girder spacings, starting from left

16	16								

4. Moments of inertia of girders, starting from left

10	10	10							

5. Torsional inertia of girders, starting from left

20	20	20							

6.

Slab thickness	E of slab material	G of slab material
1	3E6	1.5E6

7.

No. of loads in one long-line (max. 7)
1

8. Weights of loads in one longitudinal line starting from left

100					

9. Distances of loads in one longitudinal line from the left abutment

12					

10.

No. of lines of loads (max. 10)
1

11. Transverse distances of lines of loads from the outer left girder

16									

12.

No. of ref. sections (max. 10)
2

13. Distances of reference sections from the left-hand abutment

12	24								

14. Prescribed deflections at intermediate support

O	O	O							

15. Flexibility of intermediate support (in the same order as No. 14 above)

O	O	O							

16. Girder No. under which each intermediate support is located (in the same order as No. 14 above)

1	2	3							

17. Distance of each intermediate support from the left-hand abutment (in the same order as No. 14 above)

24	24	24							

Figure 9.9 Example of data input to SECAN2.

TABLE 9.12 Comparison of SECAN2 Results with Grillage Analogy

Response	Deck-slab thickness	Intermediate support for girder 1		Intermediate support for girder 2		Intermediate support for girder 3	
		Grillage	SECAN2	Grillage	SECAN2	Grillage	SECAN2
Intermediate-support reactions	1	0.6	1.9	67.6	65.0	0.6	1.9
	2	3.5	5.5	61.7	57.7	3.5	5.5
	4	9.8	10.2	49.1	48.8	9.8	10.2
Girder moments at transverse section containing load	1	21.7	24.8	443.1	437.8	21.7	24.8
	2	70.1	74.5	347.1	338.5	70.1	74.5
	4	118.5	115.3	250.4	256.9	118.5	115.3
Girder moments at transverse section containing intermediate supports	1	−4.5	−7.8	−216.0	−209.0	−4.5	−7.8
	2	−23.9	−29.1	−177.2	−166.7	−23.9	−29.1
	4	−54.2	−57.0	−116.6	−111.1	−54.2	−57.0

NOTES: 1. For bridge details see Fig. 9.8.
2. All units are compatible with dimensions given in Fig. 9.8.

the program also prints the various vectors and matrices for the different steps of calculation.

Example. To demonstrate the use of SECAN2 and to test the validity of the program, a fictitious three-girder bridge with three intermediate supports is analyzed for three slab thicknesses and the results compared with those obtained by the grillage analogy method.

Details of the bridge are given in Fig. 9.8, and the data input on the input sheet in Fig. 9.9. Figure 9.8 also shows the mesh layout for the grillage analysis. To demonstrate that the program can deal with any compatible set of units, the units are deliberately omitted in Fig. 9.8 and the rest of the example.

Intermediate-support reactions and girder moments at the cross sections containing the load and supports respectively, as obtained by the two methods, are tabulated in Table 9.12. It is encouraging to note that the maximum values of the responses obtained by SECAN2 are always within 6 percent of those given by the grillage method. For responses other than the maximum ones, the differences between the results of the two methods may appear larger percentagewise but are small in magnitude. It can be said with confidence that SECAN2, which requires far less computing power than the grillage analogy method, gives results of comparable accuracy.

It may be noted that the SECAN2 results listed in Table 9.12 were obtained by using only three harmonics. Increasing the number of harmonics to five made an imperceptible difference in the results.

Manual Method

10.1 Introduction

As shown in Sec. 4.4, the problem of slow convergence of the harmonic series can be avoided by subtracting from the free-response diagram those load effects which are distributed to other girders. This process is schematically shown in Fig. 10.1 for a four-girder bridge with a single-point load on an outer girder at the midspan. The free-moment diagram is obtained by assuming that all the load is sustained by the loaded girder. By considering only the first harmonic, the portion of the bending moment which is distributed to each of the three girders is in the shape of a half sine wave; so is also therefore the sum of the moments distributed to the three girders. If we subtract from the free-moment diagram the total moment distributed to the other three girders, we are left with the diagram of moment retained by the loaded girder.

It is instructive to note that the variation of the retained moment in the loaded girder with respect to the span is not linear, as was the free moment. This departure from linearity can be accounted for by only rigorous methods. Most of the simplified methods, e.g., Refs. 1, 2, and 3, are based on the assumption that the transverse distribution pattern of responses, viz., longitudinal moments and shears, is independent of the longitudinal position of the reference section. The im-

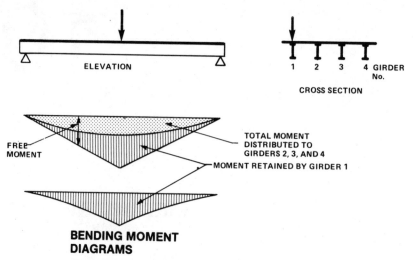

Figure 10.1 Distribution of moments in a four-girder bridge.

plication of this assumption, which leads to the linearity of the re-tained-moment diagram, is discussed in the following paragraphs.

Consider the right slab-on-girder bridge shown in Fig. 10.2. Further, consider two transverse sections, identified as 1–1 and 2–2 in the figure, away from the applied loads and at a distance Δx apart. Let the average longitudinal moment girder at 1–1 and 2–2 be M_1 and M_2 respectively. The average longitudinal shear per girder at section 0–0, which is

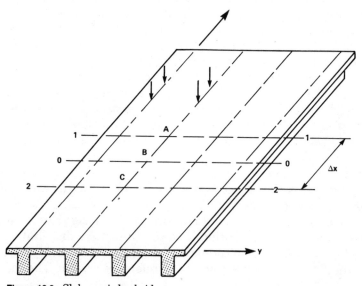

Figure 10.2 Slab-on-girder bridge.

midway between 1–1 and 2–2, is equal to V_0 and is given by

$$V_0 = \frac{M_1 - M_2}{\Delta x} \tag{10.1}$$

It is recalled that Eq. (10.1) is the finite difference equivalent of

$$V_x = \frac{dM_x}{dx} \tag{10.2}$$

Now consider points A, B, and C along a girder at sections 1–1, 0–0, and 2–2 respectively, and let the longitudinal moments at A and C be designated as M_A and M_C respectively. The above-mentioned assumption, which is the basis of several simplified methods of bridge analysis, requires that

$$\frac{M_A}{M_1} = \frac{M_C}{M_2} = K \tag{10.3}$$

where K is a constant.

If the effect of twisting moments on longitudinal vertical planes is ignored, then the intensity of longitudinal shear V_B at point B, which is midway between A and C, is given by

$$V_B = \frac{M_A - M_C}{\Delta x} \tag{10.4}$$

By substituting the expressions for M_A and M_C from Eq. (10.3), Eq. (10.4) can be rewritten as

$$V_B = K \frac{M_1 - M_2}{\Delta x} \tag{10.5}$$

or, from Eq. (10.1),

$$V_B = KV_0 \tag{10.6}$$

Equation (10.6) implies that longitudinal shear has the same transverse pattern of distribution as longitudinal moments.

It has been shown elsewhere in the book that in practice the distribution of longitudinal shears can be markedly different from that of longitudinal moments. This implies that, strictly speaking, Eq. (10.3) and hence the assumption on which it is based are false. Thus it can be stated with confidence that the variation of retained moments in a girder with respect to the distance along the span is not linear. The actual variation can be accounted for only by rigorous methods.

10.2   Development of the Manual Method

For bridges having equal and equally spaced girders, the amount of calculations required to analyze a bridge by the semicontinuum method can be considerably reduced if (1) only the first terms of the harmonic series are distributed and (2) the simplifications given below are adopted with respect to torsional effects and loads between girders. It is noted that, in spite of these simplifications, the method still remains rigorous and very accurate and can account for different distribution patterns of longitudinal moments and shears; it is then so simple that the necessary calculations can be performed manually, with the aid of charts.

Torsional effects.   For deflections, longitudinal moments, and shears, it is found by experience that the dominant torsional effect is given by the sum $(\lambda + \mu)$ of the torsional parameters in the longitudinal and transverse directions. The individual values of λ and μ, which are defined by Eq (4.39), are of comparatively little consequence provided that their sum remains the same. Thus an "overall" torsional parameter can be defined as $\beta = \lambda + \mu$. (10.7)

It may be noted that in determining deflections the predominance of the sum of the torsional rigidities, rather than the individual values of longitudinal and torsional rigidities, has its direct counterpart in the theory of orthotropic plates. In that theory, the deflection of the plate, using standard notation, is given by the solution of the equation

$$D_x \frac{\partial^4 w}{\partial x^4} + 2H \frac{\partial^4 w}{\partial x^2 \, \partial y^2} + Dy \frac{\partial^4 w}{\partial y^4} = q \tag{10.8}$$

where $2H$ is the sum of the longitudinal and torsional rigidities. For zero Poisson's ratio, $2H$ is equal to $(D_{xy} + D_{yx})$. From this it is clear that deflections respond to the sum of the torsional rigidities rather than to their individual values.

For convenience λ is set equal to zero in the $[A]$ matrix of Eq. (4.38) and all the torsional ridigity is put into the longitudinal direction by replacing μ with β. The expression for β for the first harmonic becomes

$$\beta = \frac{1}{\pi^2} \left(\frac{L}{S}\right)^2 \frac{SD_{yx} + GJ}{EI} \tag{10.9}$$

and the equation for η for the first harmonic is

$$\eta = \frac{12}{\pi^4} \left(\frac{L}{S}\right)^3 \frac{LD_y}{EI} \tag{10.10}$$

Loads between girders. It is sufficiently accurate for design purposes to replace a line of wheels situated between two girders by statically equivalent lines of loads on the girders on either side. For example, the loads on a line of wheels halfway between girders 1 and 2 would be replaced by loads of one-half of the size on each of girders 1 and 2.

Designers who wish to be still more accurate and to be rigorously correct should replace the line of wheels by fractions of it applied to all the girders. The fractions are obtained by treating the transverse medium of the bridge as a transverse beam on rigid supports and finding the support reactions. These support reactions are then applied to the individual girders and are in turn distributed by using the distribution coefficients described below. It is noted, however, that the simple statical division of the load between the adjacent girders is sufficiently accurate for all design purposes.

Distribution coefficients. For a bridge with N girders, the solution of $2N$ equations, as discussed in Sec. 4.3, yields N distribution coefficients for bending moments in the girders and N distribution coefficients for twisting moments. The latter are not used in the method given in this chapter.

Values of distribution coefficients, which depend upon the values of β and η, the number of girders in the bridge, and the location of load, can be plotted in the form of contours in the η, β space for values of η and β which cover most practical bridges. For three-, four-, and five-girder bridges the coefficients can be calculated by using the formulas for distribution coefficients given in Sec. 10.4. These expressions are obtained by the solution of Eq. (4.37). Alternatively, the values of the various distribution coefficients can be read from the distribution coefficient contours given in Figs. 10.3, 10.4, and 10.5, for bridges with three, four, and five girders respectively. In these figures, η ranges between 0.1 and 1000 and β between 0.01 and 10.0.

The distribution coefficient ρ is used with two subscripts, the first one of which refers to the girder number for which the coefficient applies and the second to the loaded-girder number. The reference system for charts of various distribution coefficients is given in Table 10.1.

It is noted that the distribution coefficients mentioned above are not to be used in the same way as in the usual and less accurate distribution coefficient methods such as, for example, that of Ref. 2. The use of these coefficients is explained in the following subsection.

Free moments and shears. As discussed in Sec. 4.4, free moments and shears are the responses in a girder caused by one line of wheels, or one-half of the design loading for one lane. In the case of a simply supported girder the free moments and shears, which are statically determinate, are denoted by M_L and V_L respectively.

Following the notation of Fig. 10.6, the free moment M_L, a beam of span L and subjected to r point loads, is analyzed into the following harmonic series.

$$M_L = \frac{2L}{\pi^2} \sum_{p=1}^{p=r} W_p \left(\sin \frac{\pi x_p}{L} \sin \frac{\pi x}{L} + \frac{1}{2^2} \sin \frac{2\pi x_p}{L} \sin \frac{2\pi x}{L} + \cdots \right)$$

(10.11)

For the first harmonic, the free moment, designated as M'_L, is thus given by

$$M'_L = K_1^{(M)} \sin \frac{\pi x}{L}$$

(10.12)

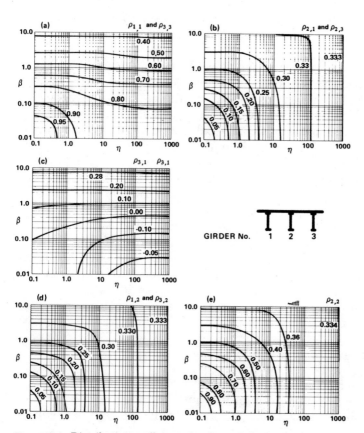

Figure 10.3 Distribution coefficients for three-girder bridges.

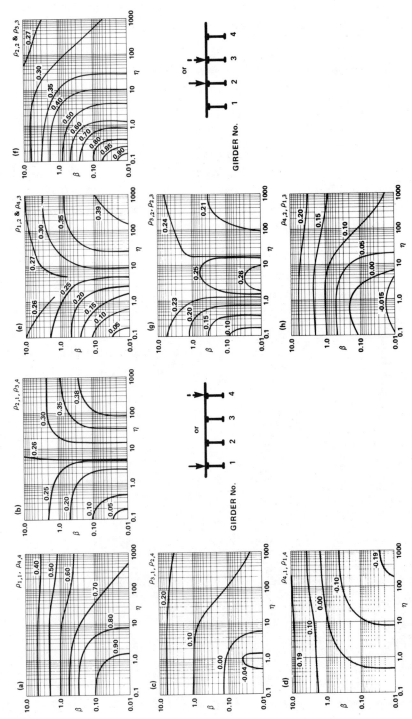

Figure 10.4 Distribution coefficients for four-girder bridges.

where
$$K_1^{(M)} = \frac{2L}{\pi^2} \sum_{p=1}^{p=r} W_p \sin \frac{\pi x_p}{L} \tag{10.13}$$

Similarly, the free shear V_L' due to the first harmonic is given by

$$V_L' = K_1^{(V)} \cos \frac{\pi x}{L} \tag{10.14}$$

where
$$K_1^{(V)} = \frac{2}{\pi} \sum_{p=1}^{p=r} W_p \sin \frac{\pi x_p}{L} \tag{10.15}$$

When the loading consists of a partially distributed load, the expressions for M_L' and V_L' are also given by Eqs. (10.12) and (10.14) respec-

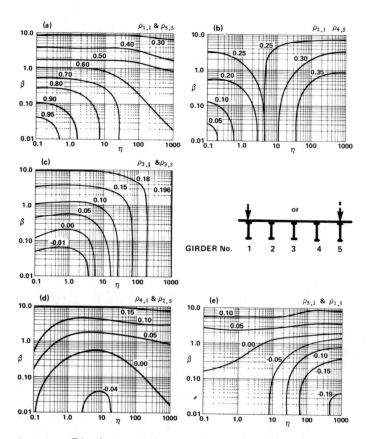

Figure 10.5 Distribution coefficients for five-girder bridges.

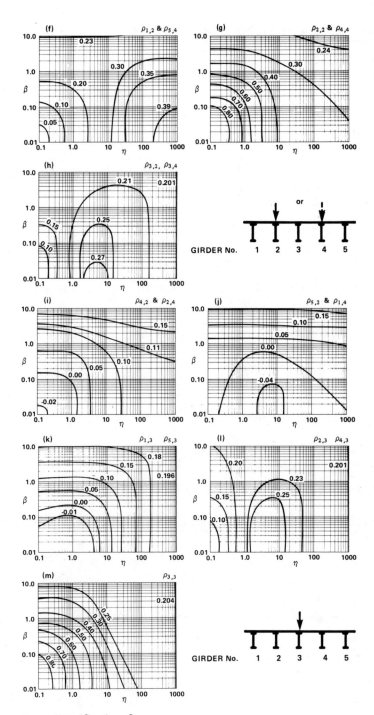

Figure 10.5 (*Continued*)

TABLE 10.1 Reference Table for Distribution Coefficient Charts

No. of girders in bridge	Load at girder no.	Coefficient at girder no.									
		1		2		3		4		5	
		Coeff.	Fig.	Coeff.	Fig.	Coeff.	Fig.	Coeff.	Fig.	Coeff.	Fig.
3	1	$\rho_{1,1}$	10.3a	$\rho_{2,1}$	10.3b	$\rho_{3,1}$	10.3c				
	2	$\rho_{1,2}$	10.3d	$\rho_{2,2}$	10.3e	$\rho_{3,2}$	10.3d				
	3	$\rho_{1,3}$	10.3c	$\rho_{2,3}$	10.3a	$\rho_{3,3}$	10.3a				
4	1	$\rho_{1,1}$	10.4a	$\rho_{2,1}$	10.4b	$\rho_{3,1}$	10.4c	$\rho_{4,1}$	10.4d		
	2	$\rho_{1,2}$	10.4e	$\rho_{2,2}$	10.4f	$\rho_{3,2}$	10.4g	$\rho_{4,2}$	10.4h		
	3	$\rho_{1,3}$	10.4h	$\rho_{2,3}$	10.4g	$\rho_{3,3}$	10.4f	$\rho_{4,3}$	10.4e		
	4	$\rho_{1,4}$	10.4d	$\rho_{2,4}$	10.4c	$\rho_{3,4}$	10.4b	$\rho_{4,4}$	10.4a		
5	1	$\rho_{1,1}$	10.5a	$\rho_{2,1}$	10.5b	$\rho_{3,1}$	10.5c	$\rho_{4,1}$	10.5d	$\rho_{5,1}$	10.5e
	2	$\rho_{1,2}$	10.5f	$\rho_{2,2}$	10.5g	$\rho_{3,2}$	10.5h	$\rho_{4,2}$	10.5i	$\rho_{5,2}$	10.5j
	3	$\rho_{1,3}$	10.5k	$\rho_{2,3}$	10.5l	$\rho_{3,3}$	10.5m	$\rho_{4,3}$	10.5l	$\rho_{5,3}$	10.5k
	4	$\rho_{1,4}$	10.5j	$\rho_{2,4}$	10.5i	$\rho_{3,4}$	10.5h	$\rho_{4,4}$	10.5g	$\rho_{5,4}$	10.5f
	5	$\rho_{1,5}$	10.5e	$\rho_{2,5}$	10.5d	$\rho_{3,5}$	10.5c	$\rho_{4,5}$	10.5b	$\rho_{5,5}$	10.5a

tively, but the values of $K_1^{(M)}$ and $K_1^{(V)}$ are obtained by the following equations.

$$K_1^{(M)} = \frac{2WL^2}{u\pi^3} \sin \frac{\pi c}{L} \sin \frac{\pi u}{L} \qquad (10.16)$$

and

$$K_1^{(V)} = \frac{2WL}{u\pi^2} \sin \frac{\pi c}{L} \sin \frac{\pi u}{L} \qquad (10.17)$$

where the notation is as defined in Fig. 10.7.

Steps of calculations. The following steps are required in the calculation of live-load moments and shears in the various girders.

1. Calculate the values of β and η from Eqs. (10.9) and (10.10) respectively.
2. As mentioned earlier, transform all lines of loads on the cross-

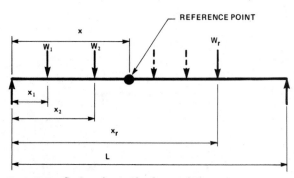

Figure 10.6 Series of point loads on a beam.

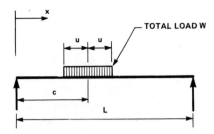

Figure 10.7 Partial uniformly distributed load on a simply supported beam.

section to statically equivalent loads on girder locations. Two ways of doing so are shown in Fig. 10.8 for one particular case. The fractions of transformed lines of load placed on girders 1 to N are designated as F_1 to F_N respectively. As a check, it should be ensured that the sum of these fractions is equal to the number of lines of loads.

3. Corresponding to the values of β and η, obtain the values of the distribution coefficients for each girder due to the load on each girder from the relevant charts of Figs. 10.3, 10.4, and 10.5.

4. Combine the effects of all transformed loads to obtain the final distribution coefficients for all girders. The value of the final distribution coefficient $\bar{\rho}_n$ for girder n is given by

$$\bar{\rho}_n = F_1\rho_{n,1} + F_2\rho_{n,2} + \cdots + F_N\rho_{n,N} \qquad (10.18)$$

5. For obtaining the retained moment M_n in girder n at a distance x from the left-hand support (see Fig. 10.6), first calculate *by statics* [not by Eq. (10.11)] the free moment M_L at the reference section due to one line of wheels, or one-half of the loading in one lane. Then

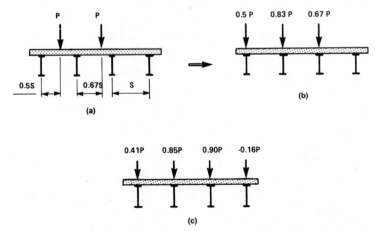

Figure 10.8 Transformation of loads to girder positions. (*a*) Actual loads. (*b*) Transformed loads by simple static transfer. (*c*) Transformed loads by rigid-support reactions.

calculate M_n by the following equation:

$$M_n = F_n M_L - K_1^{(M)}(F_n - \bar{\rho}_n) \sin \frac{\pi x}{L} \qquad (10.19)$$

where $K_1^{(M)}$ is obtained by Eq. (10.13) for a series of point loads and by Eq. (10.16) for partial uniformly distributed loads. It is recalled that the quantity $\pi x/L$ is in radians.

6. For obtaining retained shear V_n in girder n at a distance x from the left-hand support, first calculate *by statics* the free shear V_L at the reference section due to one line of wheels, or one-half of the loading in one lane. Then calculate V_n by the following equation:

$$V_n = F_n V_L - K_1^{(V)}(F_n - \bar{\rho}_n) \cos \frac{\pi x}{L} \qquad (10.20)$$

where $K_1^{(V)}$ is obtained by Eq. (10.15) for a series of point loads and by Eq. (10.17) for partial uniformly distributed loads.

10.3   Worked Example

To illustrate the use of the manual method the case of a five-girder T-beam bridge having a span of 24.0 m and subjected to an eccentrically

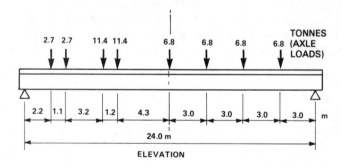

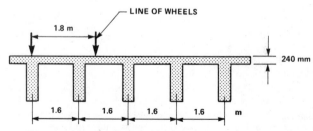

Figure 10.9 Details of loading on a bridge.

placed vehicle with eight axles is solved. The steps of the calculations are in the same order as proposed earlier.

Calculation of β and η. The cross section of the bridge and details of the loading are shown in Fig. 10.9. The flexural rigidity EI of the composite girder is found to be $2.39E$, and its torsional rigidity $11.06 \times 10^{-3}E$. The numerical quantities have the units of m^4.

For a unit length of the deck slab of thickness t, the transverse flexural rigidity D_y is equal to $Et^3/12$ and the torsional rigidity equal to $Gt^3/6$. By using these expressions,

$$D_y = 1.15 \times 10^{-3}E$$

$$D_{xy} = 1.00 \times 10^{-3}E$$

The numerical quantities have units of m^4/m. It is noted that for concrete, which has a Poisson's ratio of about 0.15, E is equal to $2.3G$.

From Eqs. (10.9) and (10.10),

$$\beta = \frac{1}{\pi^2} \left(\frac{24.0}{1.6}\right)^2 \frac{(1.6 \times 1.00) + 11.06}{2390} = 0.12$$

$$\eta = \frac{12}{\pi^4} \left(\frac{24.0}{1.6}\right)^3 \frac{24 \times 1.15}{2390} = 4.80$$

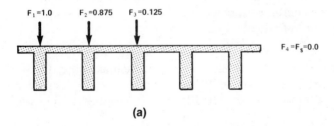

F$_1$=1.0   F$_2$=0.875   F$_3$=0.125

F$_4$=F$_5$=0.0

(a)

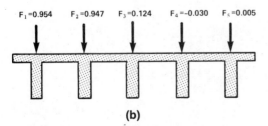

F$_1$=0.954   F$_2$=0.947   F$_3$=0.124   F$_4$=-0.030   F$_5$=0.005

(b)

Figure 10.10 Transformation of loads to girder locations. (*a*) Loads transformed to girder locations by simple static transfer. (*b*) Loads transformed to girder locations by rigid-support reactions.

Transformation of loads. As shown in Fig. 10.9, there are two lines of loads, one directly above the outer girder and the other between girders 2 and 3 at a distance of 0.2 m from girder 2. Thus, by using simple statics transfer, the fractions of the section line of load assigned to girders 2 and 3 are equal to 1.4/1.6 (= 0.875) and 0.2/1.6 (= 0.125) respectively. The transformed loads are shown in Fig. 10.10a.

If transfer of loads to girder locations is effected by rigid support reactions, the fractions of load applied to the girders are as shown in Fig. 10.10b.

Distribution coefficients. Corresponding to β equal to 0.12 and η equal to 4.80, the values of the distribution coefficients are obtained from the charts in Fig. 10.5 or from the formulas given in Sec. 10.4. These values, as obtained from the formulas, are listed in Table 10.2. As shown in this table, the sum of the coefficients for all girders due to load on any one girder is 1.000, thus confirming that no errors have been made.

It is noted that if the distribution coefficients are read from the charts, an accuracy to two decimal places rather than three is obtained, in which case the sum of distribution coefficients may turn out to be 0.99 or 1.01 owing to rounding error.

Combined distribution coefficients using simple statics transfer of loads. By using Eq. (10.18), the distribution factors obtained above are multiplied by the load fractions shown in Fig. 10.10a as shown in Table 10.3. The table also shows the combined factors $\bar{\rho}_n$, which do add up to be equal to the number of lines of wheels, i.e., 2.000.

Retained moments. Figure 10.11a shows the free-moment diagram M_L in a girder due to one line of wheels. We investigate moments in girder 2 at $x = 6.5$ and 12.0 m. From Eq. (10.13) $K_1^{(M)}$ is obtained as follows:

$$K_1^{(M)} = \frac{2 \times 24}{\pi^2} \left(1.35 \sin \frac{2.2\pi}{24} + 1.35 \sin \frac{3.3\pi}{24} + 5.7 \sin \frac{6.5\pi}{24} \right.$$

$$+ 5.7 \sin \frac{7.7\pi}{24} + 3.4 \sin \frac{12\pi}{24} + 3.4 \sin \frac{15\pi}{24}$$

$$\left. + 3.4 \sin \frac{18\pi}{24} + 3.4 \sin \frac{21\pi}{24} \right)$$

$$= 98.7 \text{ t} \cdot \text{m}$$

Thus, according to Eq. (10.19), the moment in girder 2 at x equal to

TABLE 10.2   Distribution Factors for the Five-Girder Bridge Shown in Fig. 10.9 ($\beta = 0.12$; $\eta = 4.80$)

Distribution coefficient for girder no.	Load on girder no.				
	1	2	3	4	5
1	0.786	0.257	0.011	−0.035	−0.019
2	0.257	0.447	0.264	0.067	−0.035
3	0.011	0.264	0.450	0.264	0.011
4	−0.035	0.067	0.264	0.447	0.257
5	−0.019	−0.035	0.011	0.257	0.786
	1.000	1.000	1.000	1.000	1.000

6.5 m is found as shown below:

$$(M_2)_{x=6.5\text{m}} = 0.875 \times 85.22 - 98.7(0.875 - 0.681) \sin \frac{6.5\pi}{24}$$

$$= 60.25 \text{ t} \cdot \text{m}$$

Similarly, $(M_2)_{x=12.0\text{m}}$ is found to be 64.33 t · m.

Retained shears.   Shears at x equal to 0.0 m, i.e., at the left support, are calculated here for girder 2. Free shears due to one line of wheels are shown in Fig. 10.11b. K_1^y is found from Eq. (10.15) to be equal to 12.9. Therefore, from Eq. (10.20),

$$(V_2)_{x=0.0} = 0.875 \times 14.67 - 12.9(0.875 - 0.681) \cos \frac{\pi(0)}{L} = 10.33 \text{ t}$$

Comparison of results by using the two methods of transferring loads to girder positions.   In constructing Table 10.3 the columns of Table 10.2 were multiplied by the load fractions on girders shown in Fig. 10.10a. Similarly, if one wishes to be more precise in handling the load transfer

TABLE 10.3   Combined Distribution Factors Using Simple Statics Transfer of Loads to Girder Positions

n	$F_1\rho_{n,1}$	$F_2\rho_{n,2}$	$F_3\rho_{n,3}$	$F_4\rho_{n,4}$	$F_5\rho_{n,5}$	$\bar{\rho}_n$
1	0.786	0.225	0.001	0	0	1.012
2	0.257	0.391	0.033	0	0	0.681
3	0.011	0.231	0.057	0	0	0.299
4	−0.035	0.059	0.033	0	0	0.057
5	−0.019	−0.031	0.001	0	0	−0.049
						2.000

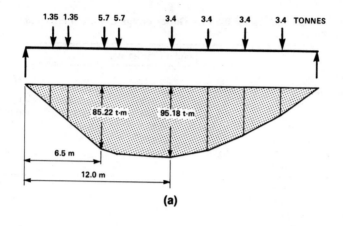

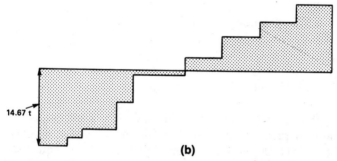

Figure 10.11 Free-moment and free-shear diagrams due to one line of wheels. (a) Free-moment diagram, M_L. (b) Free-shear diagram, V_L.

phenomenon, the columns of Table 10.2 are multiplied by the load fractions shown in Fig. 10.10b. The results are then as shown in Table 10.4.

If, for example, the bending moment in girder 2 at the longitudinal position $x = 6.5$ m is now sought, the following is obtained:

$$(M_2)_{x=6.5 \text{ m}} = 0.947 \times 85.22 - 98.7(0.947 - 0.699) \sin \frac{6.5\pi}{24}$$

$$= 62.30 \text{ t} \cdot \text{m}$$

and similarly, at $x = 12$ m,

$$(M_2)_{x=12 \text{ m}} = 0.947 \times 95.18 - 98.7(0.947 - 0.699) \sin \frac{12\pi}{24}$$

$$= 65.66 \text{ t} \cdot \text{m}$$

TABLE 10.4 **Combined Distribution Factors Using Rigid-Support Reactions for Transfer of Loads to Girder Positions**

n	$F_1\rho_{n,1}$	$F_2\rho_{n,2}$	$F_3\rho_{n,3}$	$F_4\rho_{n,4}$	$F_5\rho_{n,5}$	$\bar{\rho}_n$
1.	0.750	0.243	0.001	0.001	0.000	0.995
2.	0.245	0.423	0.033	-0.002	0.000	0.699
3.	0.011	0.250	0.056	-0.008	0.000	0.309
4.	-0.033	0.063	0.033	-0.013	0.001	0.051
5.	-0.018	-0.033	0.001	-0.008	0.004	-0.054
						2.000

It will be noted that these values differ by only about 2 t · m from those obtained by using simple statics transfer. This difference is only about 1 percent of the total maximum moment that is on the bridge due to the two lines of wheels. Also, it will be noted from an inspection of the column $\bar{\rho}_n$ in Tables 10.3 and 10.4 that the most heavily loaded girder (in this case, girder 1) is given a slightly larger fraction of the total load when simple statics transfer is used. In other words, this method of transferring loads onto girder positions is slightly conservative, which is a good thing.

The conclusion reached is that the extra precision of using the rigid-support-reaction approach is quite unnecessary. Simple statics transfer is recommended strongly.

Validity of the method. The example solved was also solved by the computer program SECAN1 discussed in Chap. 9 by using five harmonics. It is noted that virtually complete convergence can be expected after five harmonics and that the simplification regarding the apportioning of loads to girder locations is not used in this program. Various results

TABLE 10.5 **Comparison of Results by the Proposed Method with the Computer-Based Method**

	Girder 1			Girder 2		
Method	$(M_1)_{x=6.5m}$, t · m	$(M_1)_{x=12.0m}$, t · m	$(V_1)_{x=0.0m}$, t	$(M_2)_{x=6.5m}$, t · m	$(M_2)_{x=12.0m}$, t · m	$(V_2)_{x=0.0m}$, t
Proposed method	86.11	96.36	14.82	60.25	64.33	10.33
Computer-based method	84.15	94.25	14.49	62.15	66.1	10.59
Percentage difference	2.3	2.2	2.2	-3.1	-2.7	-2.6

by the two methods are compared in Table 10.5. It can be seen that the difference between the two sets of values is almost negligible, being always less than 3.1 percent. It is comforting to note that the error resulting from the manual method now presented is always on the safe side for the most heavily loaded girder.

10.4 Expressions for Distribution Coefficients

Expressions for distribution coefficients $\rho_{m,n}$ are given in the following paragraphs with respect to the girder numbering system shown in Fig. 10.12. It is again noted that the subscript m refers to the girder for which the coefficient applies and n to the loaded girder.

Three-girder bridges

1. *Load on girder 1*

$$\rho_{1,1} = \frac{2 + (\eta + 4\beta)}{4 + 3(\eta + 4\beta)} + \frac{1}{2(1 + \beta)}$$

$$\rho_{2,1} = \frac{(\eta + 4\beta)}{4 + 3(\eta + 4\beta)} \tag{10.21}$$

$$\rho_{3,1} = \frac{2 + (\eta + 4\beta)}{4 + 3(\eta + 4\beta)} - \frac{1}{2(1 + \beta)}$$

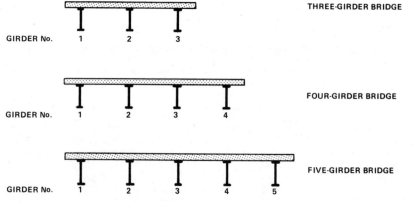

THREE-GIRDER BRIDGE

GIRDER No.   1   2   3

FOUR-GIRDER BRIDGE

GIRDER No.   1   2   3   4

FIVE-GIRDER BRIDGE

GIRDER No.   1   2   3   4   5

Figure 10.12 Cross sections of slab-on-girder bridges.

2. *Load on girder 2*

$$\rho_{1,2} = \frac{(\eta + 4\beta)}{4 + 3(\eta + 4\beta)}$$

$$\rho_{2,2} = \frac{4 + (\eta + 4\beta)}{4 + 3(\eta + 4\beta)} \qquad (10.22)$$

$$\rho_{3,2} = \rho_{1,2}$$

Four-girder bridges

1. *Load on girder 1*

$$\rho_{1,1} = \frac{0.5(5 + 5\beta + 0.5\eta)}{\Delta_1} + \frac{1.5(1 + 3\beta + 1.5\eta)}{\Delta_2}$$

$$\rho_{2,1} = \frac{0.5(5\beta + 0.5\eta)}{\Delta_1} + \frac{1.5(\beta + 0.5\eta)}{\Delta_2}$$

$$\rho_{3,1} = \frac{0.5(5\beta + 0.5\eta)}{\Delta_1} - \frac{1.5(\beta + 0.5\eta)}{\Delta_2} \qquad (10.23)$$

$$\rho_{4,1} = \frac{0.5(5 + 5\beta + 0.5\eta)}{\Delta_1} - \frac{1.5(1 + 3\beta + 1.5\eta)}{\Delta_2}$$

where

$$\Delta_1 = 5 + 10\beta + \eta \qquad (10.24)$$

$$\Delta_2 = 3 + 12\beta + 5\eta + 3\eta\beta + 6\beta^2$$

2. *Load on girder 2*

$$\rho_{1,2} = \frac{0.5(5\beta + 0.5\eta)}{\Delta_1} + \frac{1.5(\beta + 0.5\eta)}{\Delta_2}$$

$$\rho_{2,2} = \frac{0.5(5 + 5\beta + 0.5\eta)}{\Delta_1} + \frac{1.5(1 + \beta + \eta/6)}{\Delta_2}$$

$$\rho_{3,2} = \frac{0.5(5 + 5\beta + 0.5\eta)}{\Delta_1} - \frac{1.5(1 + \beta + \eta/6)}{\Delta_2} \qquad (10.25)$$

$$\rho_{4,2} = \frac{0.5(5\beta + 0.5\eta)}{\Delta_1} - \frac{1.5(\beta + 0.5\eta)}{\Delta_2}$$

where Δ_1 and Δ_2 are as given by Eq. (10.24).

Five-girder bridges

1. *Load on girder 1*

$$\rho_{1,1} = \frac{14 + 56\beta + 32\eta + 28\beta^2 + 20\eta\beta + \eta^2}{\Delta_3} + \frac{4 + 8\beta + 2\eta}{\Delta_4}$$

$$\rho_{2,1} = \frac{14\beta + 5\eta + 28\beta^2 + 20\eta\beta + \eta^2}{\Delta_3} + \frac{4\beta + \eta}{\Delta_4}$$

$$\rho_{3,1} = \frac{-6\eta + 28\beta^2 + 20\eta\beta + \eta^2}{\Delta_3} \tag{10.26}$$

$$\rho_{4,1} = \frac{14\beta + 5\eta + 28\beta^2 + 20\eta\beta + \eta^2}{\Delta_3} - \frac{4\beta + \eta}{\Delta_4}$$

$$\rho_{5,1} = \frac{14 + 56\beta + 32\eta + 28\beta^2 + 20\eta\beta + \eta^2}{\Delta_3} - \frac{4 + 8\beta + 2\eta}{\Delta_4}$$

where
$$\Delta_3 = 28 + 140\beta + 68\eta + 140\beta^2 + 100\eta\beta + 5\eta^2 \tag{10.27}$$
$$\Delta_4 = 8 + 24\beta + 5\eta + 8\beta^2 + 2\eta\beta$$

2. *Load on girder 2*

$$\rho_{1,2} = \frac{14\beta + 5\eta + 28\beta^2 + 20\eta\beta + \eta^2}{\Delta_3} + \frac{4\beta + \eta}{\Delta_4}$$

$$\rho_{2,2} = \frac{14 + 42\beta + 18\eta + 28\beta^2 + 20\eta\beta + \eta^2}{\Delta_3}$$
$$+ \frac{4 + 4\beta + 0.5\eta}{\Delta_4}$$

$$\rho_{3,2} = \frac{28\beta + 22\eta + 28\beta^2 + 20\eta\beta + \eta^2}{\Delta_3} \tag{10.28}$$

$$\rho_{4,2} = \frac{14 + 42\beta + 18\eta + 28\beta^2 + 20\eta\beta + \eta^2}{\Delta_3}$$
$$- \frac{4 + 4\beta + 0.5\eta}{\Delta_4}$$

$$\rho_{5,2} = \frac{14\beta + 5\eta + 28\beta^2 + 20\eta\beta + \eta^2}{\Delta_3} - \frac{4\beta + \eta}{\Delta_4}$$

where Δ_3 and Δ_4 are as defined by Eq. (10.27).

3. *Load on girder 3*

$$\rho_{1,3} = \frac{-6\eta + 28\beta^2 + 20\eta\beta + \eta^2}{\Delta_3}$$

$$\rho_{2,3} = \frac{28\beta + 22\eta + 28\beta^2 + 20\eta\beta + \eta^2}{\Delta_3}$$

$$\rho_{3,3} = \frac{28 + 84\beta + 36\eta + 28\beta^2 + 20\eta\beta + \eta^2}{\Delta_3} \qquad (10.24)$$

$$\rho_{4,3} = \rho_{2,3}$$

$$\rho_{5,3} = \rho_{1,3}$$

where Δ_3 is as given by the first equation of Eq. (10.27).

REFERENCES

1. Bakht, B., and Jaeger, L. G.: *Bridge Analysis Simplified,* McGraw-Hill, New York, 1985.
2. Morice, P. B., and Little, G.: *The Analysis of Right Bridge Decks Subjected to Abnormal Loading,* Report Db 11, Cement and Concrete Association, London, 1956.
3. *Ontario Highway Bridge Design Code,* Ministry of Transportation and Communications, Downsview, Ontario, 1983.

Harmonic Analysis of Beams

I.1 Representation of Loads by Harmonic Series

Various expressions are given in Chap. 2 for the representation of different types of loads by harmonic series. In this portion of the appendix, a general procedure is presented for the derivation of these expressions.

Figure I.1 shows a simply supported beam of span L carrying a distributed load whose intensity at a distance x from the left-hand support is $q(x)$. It is required to express this load in the following form:

$$q(x) = q_1 \sin \frac{\pi x}{L} + q_2 \sin \frac{2\pi x}{L} + \cdots + q_n \sin \frac{n\pi x}{L} + \cdots \quad \text{(I.1)}$$

In order to find q_n, Eq. (I.1) is multiplied throughout by $\sin (n\pi x/L)$, and the resulting equality is integrated with respect to x on both sides between limits of 0 and L; i.e.,

$$\int_0^L q(x) \sin \frac{n\pi x}{L}\, dx = q_1 \int_0^L \sin \frac{\pi x}{L} \sin \frac{n\pi x}{L}\, dx + \cdots$$

$$+ q_n \int_0^L \sin \frac{n\pi x}{L} \sin \frac{n\pi x}{L}\, dx + \cdots \quad \text{(I.2)}$$

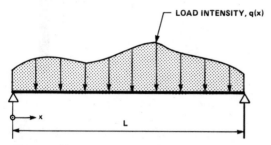

Figure I.1 Simply supported beam under a general distributed load.

It is readily shown that if m and n are two different integers, then

$$\int_0^L \sin \frac{m\pi x}{L} \sin \frac{n\pi x}{L} \, dx = 0 \tag{I.3}$$

Hence in Eq. (I.2) all integrals on the right-hand side are zero except the one incorporating q_n, so that

$$\int_0^L q(x) \sin \frac{n\pi x}{L} \, dx = q_n \int_0^L \sin^2 \frac{n\pi x}{L} \, dx \tag{I.4}$$

It is also readily shown that

$$\int_0^L \sin^2 \frac{n\pi x}{L} \, dx = \frac{L}{2} \tag{I.5}$$

Hence,

$$q_n = \frac{2}{L} \int_0^L q(x) \sin \frac{n\pi x}{L} \, dx \tag{I.6}$$

Fairly frequently the load intensity $q(x)$ is not a simple analytic function of x. However, there is no difficulty in performing the integration of Eq. (I.6) sufficiently accurately either graphically or by a numerical procedure such as Simpson's rule.

There are two particular cases of load $q(x)$, namely, uniformly distributed load and point load, which are often encountered in bridge design. Expressions for q_n are derived for these two cases in the following subsections.

Uniformly distributed load. For a uniformly distributed load, shown in Fig. I.2,

$$q(x) = q \tag{I.7}$$

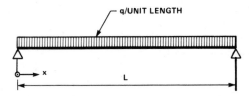

Figure I.2 Simply supported beam under a uniformly distributed load.

where q is a constant. Thus Eq. (I.6) becomes

$$q_n = \frac{2q}{L} \int_0^L \sin \frac{n\pi x}{L} \, dx \qquad (I.8)$$

whence

$$q_n = \frac{4q}{n\pi} \qquad \text{for } n \text{ odd}$$

$$= 0 \qquad \text{for } n \text{ even} \qquad (I.9)$$

giving

$$q(x) = \frac{4q}{\pi} \left(\sin \frac{\pi x}{L} + \frac{1}{3} \sin \frac{3\pi x}{L} + \cdots \right) \qquad (I.10)$$

Equation (I.10) has been used in Chap. 2 to derive Eq. (2.4a).

Point load. Figure I.3a shows a concentrated load P at a distance c from the left-hand end. This concentrated load can conveniently be regarded as a distributed load of intensity $P/2\Delta$ situated on the part of the beam between $x = (c - \Delta)$ and $x = (c + \Delta)$, as shown in Fig. I.3b, with Δ being allowed to approach zero. Then in Eq. (I.6), the intensity $q(x)$ is zero everywhere except between $x = (c - \Delta)$ and $x = (c + \Delta)$, and the equation becomes

$$q_n = \frac{2}{L} \left(\lim_{\Delta \to 0} \int_{c-\Delta}^{c+\Delta} \frac{P}{2\Delta} \sin \frac{n\pi x}{L} \, dx \right) \qquad (I.11)$$

or

$$q_n = \frac{2P}{L} \sin \frac{n\pi c}{L} \qquad (I.12)$$

This result has been used in Chap. 2 in the derivation of Eq. (2.1).

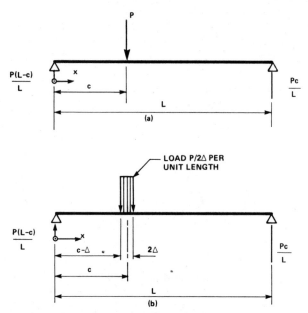

Figure I.3 Point load as a limiting case of distributed load. (*a*) Point load. (*b*) Partial uniformly distributed load.

I.2 Orthogonal-Beam Functions for Various End Conditions

It is well known that the deflection $w = w(x)$ of a uniform beam is related to the load intensity $q = q(x)$ by the equation

$$q(x) = EI \frac{d^4w(x)}{dx^4} \qquad (I.13)$$

If the load intensity and the deflection are to be of the same shape, so that $q(x) \propto w(x)$, it follows that the deflection w must satisfy a differential equation of the form

$$\frac{d^4w}{dx^4} - g^4w = 0 \qquad (I.14)$$

Then the general solution for w is

$$w = A \sin gx + B \cos gx + C \sinh gx + D \cosh gx \qquad (I.15)$$

where g is a constant whose value is yet to be determined.

Simply supported beam. For a simply supported beam of length L, the boundary conditions are $w = 0$ and $d^2w/dx^2 = 0$ at $x = 0$ and also at

$x = L$. Application of the first two conditions to Eq. (I.15) requires $B = D = 0$, and application of the remaining two then requires

$$A \sin gL + C \sinh gL = 0$$
$$-A \sin gL + C \sinh gL = 0$$

(I.16)

From Eq. (I.16), $C = 0$, since $\sinh gL \neq 0$, and then for $A \neq 0$ it is required that $\sin gL = 0$, from which

$$g = \frac{n\pi}{L} \qquad n = 1, 2, 3, \cdots$$

(I.17)

Thus, for a beam with simply supported ends, loads in the forms $\sin (\pi x/L)$, $\sin (2\pi x/L)$, ..., as in Eq. (I.1), will each result in deflections of the same shape.

Fixed-ended beam. A fixed-ended beam is now considered with the boundry conditions of $w = 0$ and $dw/dx = 0$ at $x = 0$ and also at $x = L$. Application of the first two conditions in Eq. (I.15) requires that $B + D = 0$ and $A + C = 0$, so that in this case

$$w = A(\sin gx - \sinh gx) + B(\cos gx - \cosh gx)$$

(I.18)

On now applying the remaining two boundary conditions, it is required that

$$A(\sin gL - \sinh gL) + B(\cos gL - \cosh gL) = 0$$
$$A(\cos gL - \cosh gL) + B(-\sin gL - \sinh gL) = 0$$

(I.19)

For Eq. (I.19) to be satisfied, with at least one of the constants A and B being nonzero, the determinant of their coefficients must be zero. This condition gives

$$\sinh^2 gL - \sinh^2 gL = (\cos gL - \cosh gL)^2$$

(I.20)

which simplifies to the *characteristic equation*

$$\cosh gL \cos gL = 1$$

(I.21)

Equation (I.21) has an infinite number of solutions, say, $g_1, g_2, \cdots$, where

$$g_1 = 1.5056 \frac{\pi}{L}$$

$$g_2 = 2.4998 \frac{\pi}{L}$$

(I.22)

etc.

It may be noted that the values of g_1, g_2, etc., are close to 1.5π, 2.5π, etc.

By taking the lowest value, g_1, the corresponding shape function, say $X_1(x)$, is readily obtained. Equation (I.18) is rewritten as

$$w = A_1 \left[-\frac{B_1}{A_1} (\cosh g_1 x - \cos g_1 x) - (\sinh g_1 x - \sin g_1 x) \right] \quad (I.23)$$

where the constant A_1 outside the brackets is a simple multiplier and the term inside the brackets is the shape function $X_1(x)$.

By using either equation of Eq. (I.19) it is readily found that $B_1/A_1 = -1.0178$, so that

$$X_1(x) = 1.0178(\cosh g_1 x - \cos g_1 x) - (\sinh g_1 x - \sin g_1 x) \quad (I.24)$$

Similarly,

$$X_2(x) = 0.99922(\cosh g_2 x - \cos g_2 x) - (\sinh g_2 x - \sin g_2 x) \quad (I.25)$$

and so on.

The functions $X_1(x)$, $X_2(x)$, ..., perform the same roles for the fixed-ended beam as the functions $\sin(\pi x/L)$, $\sin(2\pi x/L)$, ..., do for the simply supported beam. Figure I.4a and b shows the fixed-ended beam carrying loads in the shapes of $X_1(x)$ and $X_2(x)$ respectively, together with the bending-moment shapes $\Phi_1(x)$ and $\Phi_2(x)$ respectively. On differentiating Eqs. (I.23) and (I.24) twice each and discarding common factors of $(g_1)^2$ and $(g_2)^2$ respectively, one obtains the appropriate shape functions for bending moments:

$$\Phi_1(x) = 1.0178(\cosh g_1 x + \cos g_1 x) - (\sinh g_1 x + \sin g_1 x)$$
$$\Phi_2(x) = 0.99922(\cosh g_2 x + \cos g_2 x) - (\sinh g_2 x + \sin g_2 x) \quad (I.26)$$

It is readily shown that two such shapes are *orthogonal*; that is that, for example,

$$\int_0^L \Phi_1(x)\Phi_2(x)\, dx = 0 \quad (I.27)$$

while

$$\int_0^L [\Phi_1(x)]^2\, dx = (1.0178)^2 L = 1.0359L$$

$$\int_0^L [\Phi_2(x)]^2\, dx = (0.99922)^2 L = 0.9984L \quad (I.28)$$

etc.

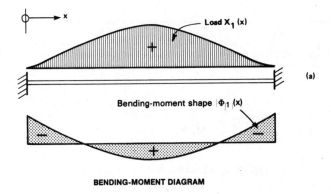

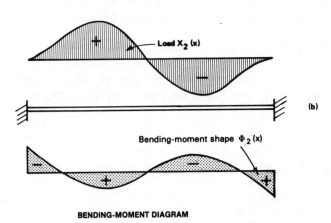

Figure I.4 Loads and bending moments in a fixed-ended beam.
(a) Loads and moments due to first harmonic. (b) Loads and
moments due to second harmonic.

By using Eqs. (I.27) and (I.28) it is easy to analyze any given bending-moment diagram, say, M_L, into a series of $\Phi(x)$ components. Let M_L be a known bending-moment diagram and write

$$M_L = M_L^{(1)}\Phi_1(x) + M_L^{(2)}\Phi_2(x) + \cdots \qquad (I.29)$$

where the amplitudes $M_L^{(1)}$, $M_L^{(2)}$, ..., are to be found.

On multiplying both sides of Eq. (I.29) by $\Phi_1(x)$ and integrating from 0 to L, the following expression is obtained:

$$\int_0^L M_L\Phi_1(x)\,dx = M_L^{(1)}\int_0^L [\Phi_1(x)]^2\,dx + M_L^{(2)}\int_0^L \Phi_1(x)\Phi_2(x)\,dx + \cdots$$

$$= (1.0359L)M_L^{(1)} + 0 + 0 + \cdots \qquad (I.30)$$

TABLE I.1   Values of $\Phi_1(x)$

x/L	0.0	0.1	0.2	0.3	0.4	0.5	0.6	0.7	0.8	0.9	1.0
$\Phi_1(x)$	-2.036	-1.094	-0.199	0.553	1.059	1.237	1.059	0.553	-0.199	-1.094	-2.036

so that

$$M_L^{(1)} = \frac{0.9653}{L} \int_0^L M_L \Phi_1(x)\, dx \qquad (I.31)$$

The integral in Eq. (I.31) is readily evaluated, for example numerically, by Simpson's rule. Normally, only three decimal places of accuracy are needed.

In order to have this integral readily evaluated on a personal computer, it is convenient to tabulate the values of $\Phi_1(x)$ from $x = 0$ to $x = L$ at intervals of $x/L = 0.1$ as shown in Table I.1. These values are stored in the program.

One then goes to the free-bending-moment diagram M_L for the externally applied loads, obtained by the usual beam-bending theory, and obtains the values of the bending moments, say, $M_0, M_1, \ldots, M_{10}$ at the same intervals of $X/L = 0.1$. For example, with the point load

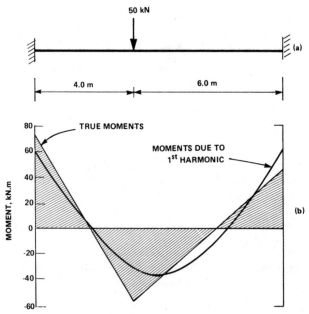

Figure I.5 Representation of bending moments of a fixed-ended beam by harmonics. (a) Load on beam. (b) Bending-moment diagrams.

TABLE I.2   Values of M_L for the Example Shown in Fig. I.5

x/L	0.0	0.1	0.2	0.3	0.4	0.5	0.6	0.7	0.8	0.9	1.0
M_L, kN	-72.0	-39.6	-7.2	25.2	57.6	40.0	22.4	4.8	-12.8	-30.4	-48.0

shown in Fig. I.5a, the free-bending-moment diagram is as shown in Fig. I.5b, and the bending moments $M_0, M_1, \ldots, M_{10}$ are as given in Table I.2.

The right-hand side of Eq. (I.31) is then quickly obtained. By multiplying each value of $\Phi_1(x)$ in turn by its associated value of M_L, Simpson's-rule multipliers $1, 4, 2, 4, \ldots, 2, 4, 1$ are then applied to these products, which are added. Recalling that the interval between successive stations is $L/10$, one then multiplies by $L/30$ and finally by the factor $0.965/L$, which appears in Eq. (I.31). The result is

$$M_L^{(1)} = -0.0655(M_0 + M_{10}) - 0.1408(M_1 + M_9) - 0.0128(M_2 + M_8)$$
$$+ 0.0712(M_3 + M_7) + 0.0681(M_4 + M_6) + 0.1592M_5 \quad (I.32)$$

Equation (I.32) is stored in the computer, and the values of M_0, $M_1, \ldots, M_{10}$ are read in as data for any particular external loading.

In the particular case of the concentrated load shown in Fig. I.5a, the insertion of moment values $M_0, M_1, \ldots,$ into Eq. (I.32) gives

$$M_L^{(1)} = 29.20 \text{ kN} \cdot \text{m} \quad (I.33)$$

On multiplying the amplitude $M_L^{(1)}$ by the values of $\Phi_1(x)$ as tabulated in Table I.1, the first harmonic of the free-bending-moment diagram M_L is obtained. Figure I.5 shows the free-bending moment M_L due to the single concentrated load and its first harmonic $29.20\Phi_1(x)$. In practice only this first harmonic is distributed to any significant degree to the unloaded girders. All the higher harmonics, involving $\Phi_2(x), \Phi_3(x)$, etc., and represented by the difference between the two curves in Fig. I.5, are retained virtually entirely by the externally loaded girder. For comparison, the values of M_L and its first harmonic component $M_L^{(1)}\Phi_1(x)$ are tabulated in Table I.3 at intervals $x/L = 0.1$.

TABLE I.3   Comparison of M_L with $M_L^{(1)}\Phi_1(x)$ for the Example Shown in Fig. I.5

x/L	0.0	0.1	0.2	0.3	0.4	0.5	0.6	0.7	0.8	0.9	1.0
M_L, kN·m	-72.0	-39.6	-7.2	25.2	57.6	40.0	22.4	4.8	-12.8	-30.4	-48.0
$M_L^{(1)}\Phi_1(x)$, kN·m	-59.5	-31.9	-5.8	16.2	30.9	36.1	30.9	16.2	-5.8	-31.9	-59.5

Replacing Twisting Moments
by Vertical Loads

The technique of replacing a distribution of twisting moments acting in the plane of a strip by intensities of vertical loading is an old one. This technique, which was developed by Thompson and Tait [1], is described in the following paragraphs.

We consider the portion of a thin horizontal beam which, as shown in Fig. II.1, is subjected to distributed twisting moments acting about horizontal axes which run perpendicular to the beam axis. The intensity of twisting moment at a distance x from a fixed datum point is denoted by T_x.

As shown in Fig. II.2, an element of beam between distances x and $(x + \delta x)$ from the datum point is subjected to a clockwise twisting moment $T_x \delta_x$. As is also shown in the figure, this twisting moment can be replaced by an upward force T_x on the left-hand edge of the element and a downward force T_x on the right.

A similar element to the immediate right of the above element is now considered. As shown in Fig. II.3, this element is subjected to a twisting moment $(T_x + \delta T_x) \delta x$, which is equivalent to an upward force $(T_x + \delta T_x)$ on the left-hand edge and a downward force $(T_x + \delta T_x)$ on the right.

The elements of Figs. II.2 and II.3 carrying the vertical loads are now put side by side along with a third element as shown in Fig. II.4.

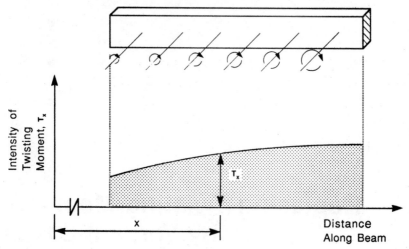

Figure II.1 Beam subjected to distributed twisting moments.

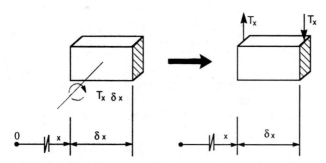

Figure II.2 Element of a beam at a distance x from the beam end.

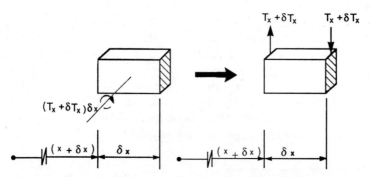

Figure II.3 Element of a beam at a distance $(x + \delta x)$ from the beam end.

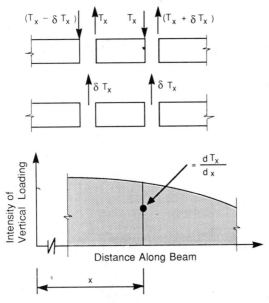

Figure II.4 Replacement of twisting-moment intensity
by upward vertical-loading intensity.

It can be seen that at the common boundary of two adjacent elements
the upward and downward forces T_x cancel each other, leaving an
upward force δT_x.

The effect is similar at the common boundaries of all elements. The
result of this operation is that the distributed clockwise twisting moments are replaced by vertically upward loading of intensity dT_x/dx
per unit length, as shown in Fig. II.4. Thus, if the twisting moment
along a beam is of intensity $A \cos (\pi x/L)$, then it can be replaced by
a distributed downward vertical force of intensity $A(\pi/L) \sin (\pi x/L)$.

REFERENCE

1. Thompson, W., and Tait, P. C.: *Treatise on Natural Philosophy,* vol. I, Cambridge,
London, 1883, part II.

Instructions for Generalized [A] Matrix and {R} Vector

The definitions for the various terms of a generalized [A] matrix and {R} vector for a bridge with N girders are given in the following. The notation used is as shown in Fig. III.1. In addition, some fictitious quantities are assumed to have the values as shown below.

$$S_0 = 0.0$$
$$b_0 = 0.0$$
(III.1)

It is noted that the quantities S_0 and b_0 do not exist in reality but are considered to exist in the abstract in order to facilitate instructions to the computer. For the same reason, it is also assumed that S_N, which does not exist in reality, is given by

$$S_N = S_{N-1}$$
(III.2)

The derivation of the equations, defined by Eqs. (4.37) and (4.38), is given in Sec. 4.3. Instructions for developing the various terms of matrix [A] and vector {R} are given in the following sections for any harmonic number. It is noted that factors k, η, λ, and μ change with every harmonic and that their values in the following equations correspond to the harmonic number under consideration.

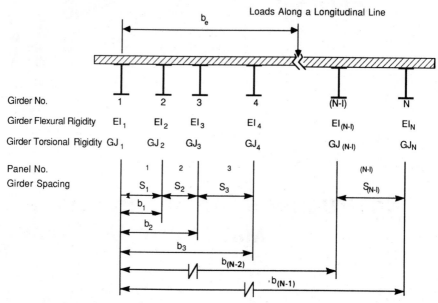

Figure III.1 Notation for the cross section of a general bridge.

III.1   Matrix [A]

The matrix $[A]$ has $(2N \times 2N)$ terms which are denoted by $A(m, n)$, where m refers to the row number and n to the column number. All the terms of matrix $[A]$ not defined below are equal to zero. It is therefore advisable to start for each harmonic by initially making all the $(2N \times 2N)$ terms of $[A]$ equal to zero.

Row 1.   The first N terms of the first row are all equal to 1.0, and the remaining terms are equal to zero; i.e.,

$$A(1, 1) = 1.0$$

$$A(1, 2) = 1.0$$

$$\vdots$$

$$A(1, N) = 1.0$$

$$A(1, N + 1) = 0.0 \qquad \text{(III.3)}$$

$$A(1, N + 2) = 0.0$$

$$\vdots$$

$$A(1, 2N) = 0.0$$

Row 2. The first N terms of the second row are given by

$$A(2, r) = \frac{1}{b_{N-1}} \left(b_{r-1} + \lambda_{r-1} \frac{k_{r-1}}{k_r} S_{r-1} - \lambda_r S_r \right) \qquad \text{(III.4)}$$

where r varies successively from 1 to N.
  The last N terms of the second row are given by

$$A(2, N + r) = \frac{1}{b_{N-1}} \mu_r S_r \qquad \text{(III.5)}$$

where r varies successively from 1 to N.

Rows 3 to (N + 1). For a row number $(2 + p)$ where p varies successively from 1 to $(N - 1)$, the various terms can be obtained as follows:
  The first term is given by:

$$A(2 + p, 1) = \left(\frac{b_p}{S_p} \right)^2 (1 + \lambda_1) - \left(\frac{b_p - b_1}{S_p} \right)^2 \lambda_1 \qquad \text{(III.6)}$$

Term t is given by Eq. (III.7), where t varies successively from 2 to p.

$$A(2 + p, t) = -\left(\frac{b_p - b_{t-2}}{S_p} \right)^2 \lambda_{t-1} \frac{k_{t-1}}{k_t} + \left(\frac{b_p - b_{t-1}}{S_p} \right)^2$$
$$\times \left(1 + \lambda_{t-1} \frac{k_{t-1}}{k_t} + \lambda_t \right) - \left(\frac{b_p - b_t}{S_p} \right)^2 \lambda_t \qquad \text{(III.7)}$$

Term $(p + 1)$ is given by

$$A(2 + p, p + 1) = -\lambda_p \frac{k_p}{k_{p+1}} \qquad \text{(III.8)}$$

Term $(N + 1)$ is given by

$$A(2 + p, N + 1) = -\frac{\eta_1}{6} \left(\frac{S_1}{S_p} \right)^2 - 2\mu_1 \left(\frac{S_1}{S_p} \right) \left(\frac{b_p}{S_p} \right) \qquad \text{(III.9)}$$

Term $(N + w)$, where w varies successively from 2 to p, is given by

$$A(2 + p, N + w) = -2\mu_w \left(\frac{S_w}{S_p} \right) \left(\frac{b_p - b_{w-1}}{S_p} \right) \qquad \text{(III.10)}$$

Term $(N + p + 1)$ is given by

$$A(2 + p, N + p + 1) = \frac{\eta_{p+1}}{6} \frac{S_{p+1}}{S_p} \qquad \text{(III.11)}$$

Rows (N + 2) to 2N. For a row ($N + 1 + d$), where d varies successively from 1 to ($N - 1$), the various terms can be obtained as follows:
The first term is given by

$$A(N + 1 + d, 1) = \left(\frac{b_d}{S_d}\right)^3 (1 + \lambda_1) - \left(\frac{b_d - b_1}{S_d}\right)^3 \lambda_1$$

$$- \frac{\eta_1}{2} \left(\frac{S_1}{S_d}\right)^3 \tag{III.12}$$

Term g is given by Eq. (III.13), where g varies successively from 2 to d.

$$A(N + 1 + d, g) = -\left(\frac{b_d - b_{g-2}}{S_d}\right)^3 \lambda_{g-1} \frac{k_{g-1}}{k_g} + \left(\frac{b_d - b_{g-1}}{S_d}\right)^3$$

$$\times \left(1 + \lambda_{g-1} \frac{k_{g-1}}{k_g} + \lambda_g\right) - \left(\frac{b_d - b_g}{S_d}\right)^3 \lambda_g \tag{III.13}$$

$$A(N + 1 + d, d + 1) = -\lambda_d \frac{k_d}{k_{d+1}} + \frac{\eta_{d+1}}{2} \left\{\frac{S_{d+1}}{S_d}\right\}^3 \tag{III.14}$$

Term ($N + 1$) is given by

$$A(N + 1 + d, N + 1) = -\frac{\eta_1}{2} \left(\frac{S_1}{S_d}\right)^2 \frac{b_d}{S_d} - 3\mu_1 \frac{S_1}{S_d} \left(\frac{b_d}{S_d}\right)^2 \tag{III.15}$$

Term ($N + f$), where f varies successively from 2 to d, is given by

$$A(N + 1 + d, N + f) = -3\mu_f \frac{S_f}{S_d} \left(\frac{b_d - b_{f-1}}{S_d}\right)^2 \tag{III.16}$$

$$A(N + 1 + d, N + 1 + d) = \frac{\eta_{d+1}}{2} \left(\frac{S_{d+1}}{S_d}\right)^3 \tag{III.17}$$

It is emphasized again that the computer program should begin by defining a $2N \times 2N$ matrix and setting all elements equal to zero. In this way, all the terms not defined above will remain equal to zero as required.

III.2   Vector {R}

The terms of vector {R} are denoted by $R(m)$, where m varies from 1 to $2N$. For a given set of loads along a longitudinal line, as shown in

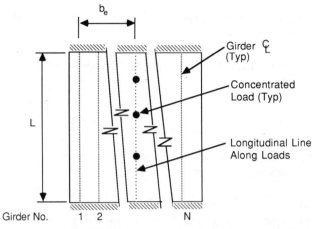

Figure III.2 Plan of a bridge showing loads on a longitudinal line.

Fig. III.2, the various terms of the vector can be obtained as follows:

Terms 1 and 2. The first and second terms are respectively equal to 1 and (b_e/b_{N-1}); i.e.,

$$R(1) = 1.0 \qquad (III.18)$$

$$R(2) = \frac{b_e}{b_{N-1}} \qquad (III.19)$$

Terms 3 to (N + 1). The term p, where p varies successively from 3 to $(N + 1)$, is given by

$$R(p) = \left[\frac{b_p - b_e}{S_p} \right]^2 \qquad (III.20)$$

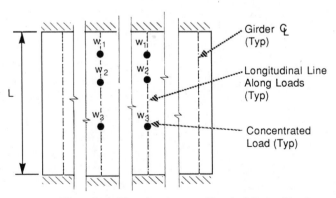

Figure III.3 Plan of a bridge showing two identical lines of loads.

The right-hand side of Eq. (III.20) is set equal to zero if the contents within the brackets are negative.

Terms (N + 2) to 2N.   The term $(N + 1 + d)$, where d varies successively from 1 to $(N - 1)$, is given by

$$R(N + 1 + d) = \left[\frac{b_d - b_e}{S_d} \right]^3 \tag{III.21}$$

The right-hand side of Eq. (III.21) is again set equal to zero if the contents within the brackets are negative.

Multiple lines of loads.   If the number of loads, their intensities, and their longitudinal position along a longitudinal line are respectively identical to those of another set of loads along a different longitudinal line, then the two sets of loads are considered to be identical in the context of semicontinuum analysis. An example of two identical sets of loads is shown in Fig. III.3.

When the sets of loads on different longitudinal lines are identical, the various $\{R\}$ vectors corresponding to each set of loads can be added together. The $\{\rho\}$ vector obtained by using the combined $\{R\}$ vector will correspond to the combined effect of all lines of loads.

Derivation of Equivalent Moment
of Inertia

This appendix presents a simple procedure by which an equivalent moment of inertia can be found for a simply supported beam of varying moment of inertia. It should be noted that, strictly speaking, this equivalent moment of inertia is for use only in the determination of transverse load distribution. However, as demonstrated later, it could also be used as a ready means of determining approximate values of beam deflections.

The following procedures are based on the assumption that, while the moment of inertia of a beam varies along the span, the modulus of elasticity of its material remains constant.

IV.1   Beams with Continuously Varying
Moment of Inertia

For derivation of the equivalent moment of inertia, a simply supported beam of span L with arbitrarily but continuously varying moment of inertia, as shown in Fig. IV.1, is considered.

The moment of inertia of the beam at a distance x from the left-hand support is denoted by I_x. This value can be represented by the following harmonic series.

$$I_x = I^{(1)} \sin \frac{\pi x}{L} + I^{(2)} \sin \frac{2\pi x}{L} + I^{(3)} \sin \frac{3\pi x}{L} + \cdots \qquad \text{(IV.1)}$$

By multiplying both sides of Eq. (IV.1) by $\sin (\pi x/L)$ and integrating with respect to x from 0 to L, the following equation is obtained.

$$\int_0^L I_x \sin \frac{\pi x}{L} \, dx = I^{(1)} \frac{L}{2} \qquad \text{(IV.2)}$$

It is noted that $\int_0^L \sin (n\pi x/L) \sin (\pi x/L) \, dx$ is equal to zero for all values of integer n except 1. Accordingly, the terms containing $I^{(2)}, I^{(3)}$, etc., on the right-hand side of Eq. (IV.2) are all equal to zero.

Equation (IV.2) can be rewritten as

$$I^{(1)} = \frac{2}{L} \int_0^L I_x \sin \frac{\pi x}{L} \, dx \qquad \text{(IV.3)}$$

It is now assumed that the behavior of the beam with varying moment of inertia shown in Fig. IV.1 can be closely represented by an equivalent beam of constant moment of inertia I_e. This latter beam is shown in Fig. IV.2. By using the same logic as for the case of a uniformly distributed load on a simply supported beam, which is dealt with in App. I, it can be shown that the moment of inertia I_x in the equivalent beam

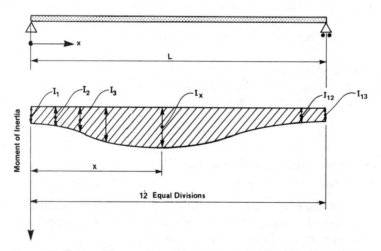

Figure IV.1 Beam with a continuously varying moment of inertia.

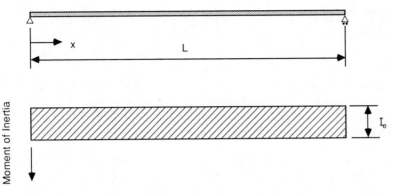

Figure IV.2 Equivalent beam with a constant moment of inertia.

is given by

$$\bar{I}_x = \frac{4I_e}{\pi}\left(\sin\frac{\pi x}{L} + \sin\frac{2\pi x}{L} + \cdots\right) \qquad (IV.4)$$

By multiplying both sides of Eq. (IV.4) by $\sin(\pi x/L)$ and integrating with respect to x from 0 to L, the following equation, similar to Eq. (IV.2), is obtained.

$$\int_0^L \bar{I}_x \sin\frac{\pi x}{L}\, dx = \frac{2I_e L}{\pi} \qquad (IV.5)$$

An assumption is now made that the beam with the constant moment of inertia can closely represent the beam with the varying moment of inertia when the first harmonic components of the moments of inertia of the two beams are equal, i.e.,

$$I^{(1)}\sin\frac{\pi x}{L} = \frac{4I_e}{\pi}\sin\frac{\pi x}{L} \qquad (IV.6)$$

or
$$I_e = \frac{I^{(1)}\pi}{4} \qquad (IV.7)$$

or, by using Eq. (IV.2),

$$I_e = \frac{\pi}{2L}\int_0^L I_x \sin\frac{\pi x}{L}\, dx \qquad (IV.8)$$

The equivalent moment of inertia I_e can now readily be obtained after determining the value of the integral $\int_0^L I_x \sin(\pi x/L)\, dx$. This integral can be conveniently handled by Simpson's rule. The beam of

Fig. IV.1 is divided into 12 equal parts, it being noted that the variation of most real-life beams can be realistically represented by this many divisions. The procedure given below can, of course, also be adapted to any convenient number of divisions.

The values of the moment of inertia at the two ends of the simply supported beam and 11 intermediate points are denoted by $I_1, I_2, \ldots,$ I_{12} and I_{13}. By using this notation, which is illustrated in Fig. IV.1, integration by Simpson's rule can be expressed as

$$\int_0^L I_x \sin \frac{\pi x}{L} \, dx = \frac{L}{36} \left[I_1 \sin (0) + 4I_2 \sin \frac{\pi}{12} + 2I_3 \sin \frac{\pi}{6} \right.$$

$$\left. + \cdots + 4I_{12} \sin \frac{11\pi}{12} + I_{13} \sin \pi \right] \quad \text{(IV.9)}$$

The various steps of calculations are listed in Table IV.1. By using these the expression for I_e for the beam shown in Fig. IV.1 can be written as

$$I_e = \frac{\pi}{72} [1.0352(I_2 + I_{12}) + 1.0000(I_3 + I_{11}) + 2.8284(I_4 + I_{10})$$

$$+ 1.7320(I_5 + I_4) + 3.8636(I_6 + I_8) + 2.0000I_7] \quad \text{(IV.10)}$$

The equivalent constant moment of inertia I_e can thus be determined by Eq. (IV.10) after the values of $I_2, I_3, \ldots, I_{12}$ have been identified.

IV.2 Beams with Symmetrical Stepped Moment of Inertia

The specific case of a beam with stepped moment of inertia symmetrical about the beam centerline is considered here. Such a case can be encountered in practice in two-span-plate girder bridges. As shown in Fig. IV.3, the girders of such bridges may be of constant depth. Their moment of inertia, however, may change in steps owing to changes in the flange cross sections.

To define the notation for this case, a beam with three steps in its moment of inertia is considered. Such a beam, together with the relevant notation, is shown in Fig. IV.4. As shown in this figure, the moment of inertia of the end segments is denoted by I, the difference between the moments of inertia of these and the adjacent segments as ΔI_1, and so on. The distances from the left support of the points where the moment of inertia changes are denoted by a_1, a_2, etc. For the case

TABLE IV.1 Steps of Calculation in Integration by Simpson's Rule

x	0	L/12	L/6	L/4	L/3	5L/12	L/2	7L/12	2L/3	3L/4	5L/6	11L/12	L
I_x	I_1	I_2	I_3	I_4	I_5	I_6	I_7	I_8	I_9	I_{10}	I_{11}	I_{12}	I_{13}
$\sin \frac{\pi x}{L}$	0.000	0.2588	0.5000	0.7071	0.8660	0.9659	1.0000	0.9659	0.8660	0.7071	0.5000	0.2588	0.0000
Simpson's multiplier, F	1	4	2	4	2	4	2	4	2	4	2	4	1
$F \sin \frac{\pi x}{L}$	0.0000	1.0352	1.0000	2.8284	1.7320	3.8636	2.0000	3.8636	1.7320	2.8284	1.0000	1.0352	0.0000

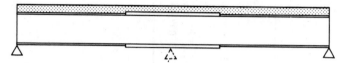

Figure IV.3 Elevation of a girder of a two-span bridge.

under consideration, because of symmetry it is clear that

$$\int_0^L I_x \sin \frac{\pi x}{L} \, dx = 2 \int_0^{L/2} I_x \sin \frac{\pi x}{L} \, dx \qquad \text{(IV.11)}$$

or

$$\int_0^L I_x \sin \frac{\pi x}{L} \, dx = 2 \left[\int_0^{a_1} I \sin \frac{\pi x}{L} \, dx \right.$$

$$+ \int_{a_1}^{a_2} (I + \Delta I_1) \sin \frac{\pi x}{L} \, dx$$

$$+ \int_{a_2}^{a_3} (I + \Delta I_1 + \Delta I_2) \sin \frac{\pi x}{L} \, dx$$

$$\left. + \int_{a_3}^{L/2} (I + \Delta I_1 + \Delta I_2 + \Delta I_3) \sin \frac{\pi x}{L} \, dx \right] \qquad \text{(IV.12)}$$

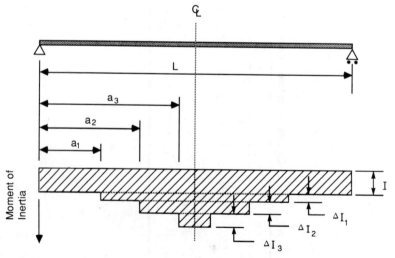

Figure IV.4 Notation for a beam with a symmetrical stepped moment of inertia.

or

$$\int_0^L I_x \sin \frac{\pi x}{L} \, dx = \frac{2L}{\pi} \left[I \left(1 - \cos \frac{\pi a_1}{L} \right) \right.$$

$$+ (I + \Delta I_1) \times \left(\cos \frac{\pi a_1}{L} - \cos \frac{\pi a_2}{L} \right)$$

$$+ (I + \Delta I_1 + \Delta I_2) \times \left(\cos \frac{\pi a_2}{L} - \cos \frac{\pi a_3}{L} \right)$$

$$+ \left. (I + \Delta I_1 + \Delta I_2 + \Delta I_3) \times \left(\cos \frac{\pi a_3}{L} \right) \right]$$

$$= \frac{2L}{\pi} \left(I + \Delta I_1 \cos \frac{\pi a_1}{L} + \Delta I_2 \cos \frac{\pi a_2}{L} \right.$$

$$\left. + \Delta I_3 \cos \frac{\pi a_3}{L} \right) \qquad \text{(IV.13)}$$

By using Eqs. (IV.2), (IV.7), and (IV.13), the following expression can be developed for I_e:

$$I_e = I + \Delta I_1 \cos \frac{\pi a_1}{L} + \Delta I_2 \cos \frac{\pi a_2}{L} + \Delta I_3 \cos \frac{\pi a_3}{L} \qquad \text{(IV.14)}$$

The generalized form of Eq. (IV.14) for a beam with n steps in the moment of inertia thus becomes

$$I_e = I + \sum_{r=1}^{r=n} \Delta I_r \cos \frac{\pi a_r}{L} \qquad \text{(IV.15)}$$

IV.3   Validity of Proposed Expressions

The validity of Eq. (IV.10) in obtaining the equivalent constant moment of inertia for beams of varying moment of inertia I_x is established in the following paragraphs by considering three specific examples. As shown in Fig. IV.5a and b, these beams are simply supported and have their moments of inertia varying in different patterns. The first pattern of variation is sinusoidal and is clearly fictitious. The other two patterns, however, correspond fairly closely to those in typical three-span continuous bridges. The actual values of span length and moments of inertia used in the examples are indeed fictitious.

The equivalent constant moment of inertia I_e is calculated by Eq.

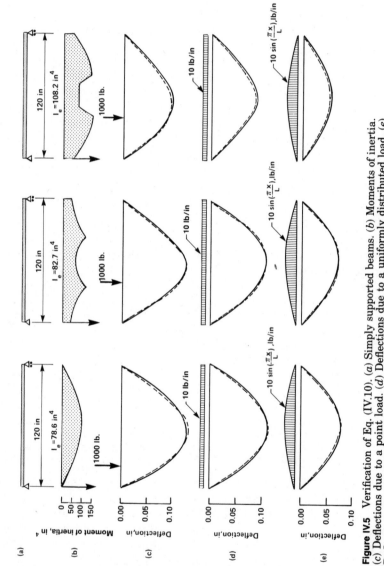

Figure IV.5 Verification of Eq. (IV.10). (a) Simply supported beams. (b) Moments of inertia. (c) Deflections due to a point load. (d) Deflections due to a uniformly distributed load. (e) Deflections due to sinusoidal load. (———— = deflection of beam with varying moment of inertia; ———— = deflection of beam with constant moment of inertia.)

(IV.10) for each of the three beams and is noted in Fig. IV.5. The beams with varying I_x and constant I_e were rigorously analyzed under different loads by the simple beam theory. The load cases included a single-point load at different positions along the span, a uniformly distributed load, and a sinusoidally distributed load. It was found that for all load cases the differences in the actual beam deflections and the corresponding deflections of the equivalent beam were extremely small. Some of the comparisons are shown graphically in Fig. IV.5c, d, and e, in which it can be observed that the deflections of the equivalent beam with constant moment of inertia are so close to the actual beam deflections that to show them apart is difficult on the scale of the figure. It is thus confirmed that a beam of constant moment of inertia, which is obtained from Eq. (IV.10), can realistically represent the behavior of a simply supported beam with varying moment of inertia under any loading.

The calculation of deflections of beams of varying moments of inertia is not always easy. By using Eq. (IV.10), and analyzing a complex beam as one of uniform moment of inertia, this task can be considerably simplified. It is noted that the harmonic method of beam analysis, discussed in Chap. 2, can be utilized for the analysis of beams with uniform moment of inertia.

The specific case of a beam shown in Fig. IV.6 is used to demonstrate the validity of Eq. (IV.15). As shown in the figure, for the quarter span next to each support the moment of inertia of the beam is I and that for the middle half span is $1.4I$. If I_e is to be calculated by using Eq. (IV.10), then the variation of the moment of inertia of the beam, using the notation of Fig. IV.1, is defined as follows:

$$I_2 = I_{12} = I$$

$$I_3 = I_{11} = I$$

$$I_4 = I_{10} = \frac{I + 1.4I}{2} = 1.2I$$

$$I_5 = I_9 = 1.4I$$

$$I_6 = I_8 = 1.4I$$

$$I_7 = 1.4I$$

By substituting the above values in Eq. (IV.10),

$$I_e = 1.2796I$$

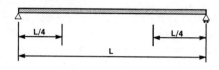

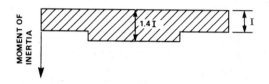

Figure IV.6 Beam with stepped moment of inertia.

By using Eq. (IV.15), the equivalent moment of inertia is given by

$$I_e = I + 0.4I \cos \frac{\pi L}{4L}$$

$$= 1.2828I$$

The value of I_e given by Eq. (IV.15) is within 0.25 percent of that given by Eq. (IV.10), thus confirming the validity of the former equation.

Listings of CONBIM

This program analyzes multispan beams by the harmonic method.

```
C
C   CONBIM
C   PROGRAM TO ANALYZE MULTI-SPAN BEAMS BY HARMONIC METHOD
C
      DIMENSION XLOAD(10),PLOAD(10),XCOL(4),DD(4),PD(4),
     *            A(4,4),D(4),B(4),XREF(20),COUNT(4),
     *            RCT(2),SHR(20),MOM(20),XREFW(20),SUMW(20)
      REAL MOM
      INTEGER COUNT,TEST
C   DATA INPUT
      READ(5,*) SPAN,EI,NLOAD,NCOL,NREF,NREFW
      READ(5,*) (XLOAD(I),I=1,NLOAD)
      READ(5,*) (PLOAD(I),I=1,NLOAD)
      READ(5,*) (XCOL(I),I=1,NCOL)
      READ(5,*) (XREF(I),I=1,NREF)
      IF(NREFW.EQ.0) GOTO 10
      READ(5,*) (XREFW(I),I=1,NREFW)
C   ECHO PRINTING OF DATA
   10 WRITE(6,500)
      WRITE(6,600) SPAN,EI
      WRITE(6,700)
      DO 20 I=1,NCOL
      COUNT(I)=0
      PD(I)=100.
   20 WRITE(6,800) I,XCOL(I)
      WRITE(6,900)
      DO 30 I=1,NLOAD
   30 WRITE(6,1000) PLOAD(I),XLOAD(I)
      WRITE(6,1100)
      NUM=3*NCOL
C
      DO 70 N=1,52
      IF(N.EQ.51) GOTO 2100
      CALL AMATRX(NCOL,SPAN,EI,N,XCOL,A)
      CALL DVECT(SPAN,EI,N,NCOL,NLOAD,XCOL,XLOAD,PLOAD,W,D)
      CALL EQN(A,NCOL,D)
      TEST=0
      DO 50 I=1,NCOL
      IF(D(I).EQ.0.) GOTO 50
      DD(I)=ABS((D(I)-PD(I))*100./D(I))
      IF(DD(I).LE.0.1) GOTO 40
      COUNT(I)=0
      GOTO 50
   40 COUNT(I)=COUNT(I)+1
      IF(COUNT(I).GE.3) COUNT(I)=3
   50 TEST=TEST+COUNT(I)
      IF(TEST.EQ.NUM) GOTO 80
      DO 60 I=1,NCOL
   60 PD(I)=D(I)
   70 CONTINUE
C
   80 WRITE(6,1400) N
      WRITE(6,1500)
      DO 90 I=1,NCOL
   90 WRITE(6,1600) I,D(I)
      CALL REACT(SPAN,NCOL,XCOL,NLOAD,XLOAD,PLOAD,D,RCT)
      CALL SHRAMO(NCOL,XCOL,NLOAD,XLOAD,PLOAD,NREF,XREF,D,RCT,SHR,MOM)
      IF(NREFW.EQ.0) GOTO 100
```

```
      CALL REFW(SPAN,EI,N,NLOAD,XLOAD,PLOAD,NCOL,XCOL,
     *          D,NREFW,XREFW,SUMW)
  100 WRITE(6,1200)
      WRITE(6,1300) RCT(1),RCT(2)
      WRITE(6,1700)
      DO 110 I=1,NREF
  110 WRITE(6,1800) XREF(I),SHR(I),MOM(I)
      IF(NREF.EQ.0) GOTO 130
      WRITE(6,1900)
      DO 120 I=1,NREFW
  120 WRITE(6,2000) XREFW(I),SUMW(I)
  130 CONTINUE
  500 FORMAT(1X,'INPUT DATA',/)
  600 FORMAT(1X,'SPAN = ',F5.1,8X,'EI = ',E9.3)
  700 FORMAT(/,1X,'COLUMN',5X,'DIST. FROM LEFT')
  800 FORMAT(3X,I1,12X,F5.1)
  900 FORMAT(/,1X,'LOAD',7X,'DIST. FROM LEFT')
 1000 FORMAT(F8.0,8X,F5.1)
 1100 FORMAT(/,35('*'))
 1200 FORMAT(/,1X,'LEFT END REACTION',5X,'RIGHT END REACTION')
 1300 FORMAT(5X,F8.2,13X,F9.2)
 1400 FORMAT(/,1X,I2,' HARMONICS REQUIRED FOR CONVERGENCE')
 1500 FORMAT(/,1X,'COLUMN',5X,'REACTION')
 1600 FORMAT(3X,I1,7X,F9.2)
 1700 FORMAT(/,1X,'DIST. FROM LEFT',8X,'SHEAR',9X,'MOMENT')
 1800 FORMAT(6X,F6.2,9X,F9.2,5X,F9.2)
 1900 FORMAT(/,1X,'DIST. FROM LEFT',8X,'DEFLECTION')
 2000 FORMAT(6X,F6.2,13X,E9.3)
 2100 IF(N.LT.51) GOTO 140
      WRITE(6,2200)
 2200 FORMAT(////,1X,'TOO MANY HARMONICS - PROGRAM STOPPED')
  140 STOP
      END
C
C   SUBROUTINE TO CALCULATE THE A MATRIX
C
      SUBROUTINE AMATRX(NCOL,SPAN,EI,N,XCOL,A)
      DIMENSION A(4,4),XCOL(4)
      DO 10 I=1,NCOL
      DO 10 J=1,NCOL
   10 A(I,J)=0.
      P=1.0
      DO 20 I=1,NCOL
      DO 20 J=1,NCOL
      C=XCOL(I)
      X=XCOL(J)
      CALL DEFLCT(SPAN,C,X,EI,P,N,W)
      A(I,J)=W
      IF(I.EQ.J) A(I,J)=A(I,J)
   20 CONTINUE
      RETURN
      END
C
C   SUBROUTINE TO CALCULATE THE D VECTOR
C
      SUBROUTINE DVECT(SPAN,EI,N,NCOL,NLOAD,XCOL,XLOAD,PLOAD,W,D)
      DIMENSION D(4),XCOL(4),XLOAD(10),PLOAD(10)
      DO 10 I=1,NCOL
```

```
 10 D(I)=0.
    DO 20 I=1,NCOL
    C=XCOL(I)
    DO 20 J=1,NLOAD
    X=XLOAD(J)
    P=PLOAD(J)
    CALL DEFLCT(SPAN,C,X,EI,P,N,W)
 20 D(I)=D(I)+W
    RETURN
    END
C
C   SUBROUTINE TO SOLVE EQUATIONS
C
    SUBROUTINE EQN(A,N,B)
    DIMENSION A(4,4),B(4),II(4),INDEX(4,2),P(4)
    INTEGER ROW,COL
    DET=1.0
    DO 10 J=1,N
 10 II(J)=0.
    DO 130 I=1,N
    T=0.
    DO 60 J=1,N
    IF(II(J)-1) 20,60,20
 20 DO 50 K=1,N
    IF(II(K)-1) 30,50,140
 30 IF(ABS(T)-ABS(A(J,K))) 40,50,50
 40 ROW=J
    COL=K
    T=A(J,K)
 50 CONTINUE
 60 CONTINUE
    II(COL)=II(COL)+1.0
    IF(ROW-COL) 70,90,70
 70 DET=-DET
    DO 80 L=1,N
    T=A(ROW,L)
    A(ROW,L)=A(COL,L)
 80 A(COL,L)=T
    T=B(ROW)
    B(ROW)=B(COL)
    B(COL)=T
 90 INDEX(I,1)=ROW
    INDEX(I,2)=COL
    P(I)=A(COL,COL)
    DET=DET*P(I)
    A(COL,COL)=1.
    DO 100 L=1,N
100 A(COL,L)=A(COL,L)/P(I)
    B(COL)=B(COL)/P(I)
    DO 130 LI=1,N
    IF(LI-COL) 110,130,110
110 T=A(LI,COL)
    A(LI,COL)=0.
    DO 120 L=1,N
120 A(LI,L)=A(LI,L)-A(COL,L)*T
    B(LI)=B(LI)-B(COL)*T
130 CONTINUE
140 RETURN
    END
C
```

```
C    SUBROUTINE TO CALCULATE DEFLECTION DUE TO A SINGLE LOAD
C
      SUBROUTINE DEFLCT(SPAN,C,X,EI,P,N,W)
      C1=2*P*(SPAN**3)/(EI*97.409083)
      C2=3.1415926*C/SPAN
      C3=3.1415926*X/SPAN
      W=0.
      DO 10 I=1,N
   10 W=W+C1*SIN(I*C2)*SIN(I*C3)/(I**4)
      RETURN
      END
C
C    SUBROUTINE TO CALCULATE END REACTIONS
C
      SUBROUTINE REACT(SPAN,NCOL,XCOL,NLOAD,XLOAD,PLOAD,D,RCT)
      DIMENSION XCOL(4),D(4),XLOAD(10),PLOAD(10),RCT(2)
      RM=0.
      SUMLD=0.
      SUMD=0.
      DO 10 I=1,NCOL
      RM=RM+XCOL(I)*D(I)
   10 SUMD=SUMD+D(I)
      DO 20 I=1,NLOAD
      RM=RM-XLOAD(I)*PLOAD(I)
   20 SUMLD=SUMLD+PLOAD(I)
      RCT(2)=-RM/SPAN
      RCT(1)=SUMLD-SUMD-RCT(2)
      RETURN
      END
C
C    SUBROUTINE TO CALCULATE THE SHEARS AND MOMENTS
C
      SUBROUTINE SHRAMO(NCOL,XCOL,NLOAD,XLOAD,PLOAD,NREF,XREF,
     *              D,RCT,SHR,MOM)
      DIMENSION SHR(20),MOM(20),RCT(2),XCOL(4),XREF(20),D(4),
     *          XLOAD(10),PLOAD(10)
      REAL MOM
      DO 40 I=1,NREF
      SHR(I)=RCT(1)
      MOM(I)=RCT(1)*XREF(I)
      DO 10 J=1,NCOL
      IF(XCOL(J).GE.XREF(I)) GOTO 20
      SHR(I)=SHR(I)+D(J)
   10 MOM(I)=MOM(I)+D(J)*(XREF(I)-XCOL(J))
   20 DO 30 J=1,NLOAD
      IF(XLOAD(J).GE.XREF(I)) GOTO 40
      SHR(I)=SHR(I)-PLOAD(J)
   30 MOM(I)=MOM(I)-PLOAD(J)*(XREF(I)-XLOAD(J))
   40 CONTINUE
      RETURN
      END
C
C    SUBROUTINE TO CALCULATE ADD DEFLECTIONS DUE TO VARIOUS SINGLE LOADS
C
      SUBROUTINE REFW(SPAN,EI,N,NLOAD,XLOAD,PLOAD,NCOL,XCOL,
     *              D,NREFW,XREFW,SUMW)
      DIMENSION XLOAD(10),PLOAD(10),XCOL(4),D(4),XREFW(20),SUMW(20)
      DO 30 I=1,NREFW
      SUMW(I)=0.
      C=XREFW(I)
```

```
      DO 10 J=1,NLOAD
      P=PLOAD(J)
      X=XLOAD(J)
      CALL DEFLCT(SPAN,C,X,EI,P,N,W)
   10 SUMW(I)=SUMW(I)+W
      DO 20 K=1,NCOL
      P=-D(K)
      X=XCOL(K)
      CALL DEFLCT(SPAN,C,X,EI,P,N,W)
   20 SUMW(I)=SUMW(I)+W
   30 CONTINUE
      RETURN
      END
```

VI

Listings of SECAN1

This program analyzes right simply supported bridges.

```
C
C     SECAN1
C     PROGRAM TO ANALYSE RIGHT, SIMPLY SUPPORTED BRIDGES
C
      DIMENSION GS(10),GMI(10),GTI(10),
     *DLS(7),DG(10),XX(20),DLG(10),W(7)
      DIMENSION AM(20,20),BM(5),RM(20),
     *KT(10),MLC(10),C(10),LAMDA(10),ALFA(10),MU(10),NUMB(10)
      DIMENSION XREF(10),HARC(5,20),ABM(10,10),AS(10,10),CINF(20)
      REAL KT,MLC,MU,LAMDA
      CHARACTER*52 TITLE
      READ(5,510) KZ,TITLE
C     1 MAKE KZ=2 IF INTERMEDIATE RESULTS ARE REQUIRED,
C     2 OTHERWISE MAKE IT EQUAL TO 1
      READ(5,*) N,NG,SPAN,E,G
      NGG=NG-1
      READ(5,*) (GS(I),I=1,NGG)
      READ(5,*) (GMI(I),I=1,NG)
      READ(5,*) (GTI(I),I=1,NG)
      READ(5,*) T,EC,GC
      READ(5,*) M
      READ(5,*) (W(I),I=1,M)
      READ(5,*) (DLS(J),J=1,M)
      READ(5,*) NW
      READ(5,*) (DLG(I),I=1,NW)
      READ(5,*) NREF
      READ(5,*) (XREF(I),I=1,NREF)
      WRITE(6,520)
      WRITE(6,530) TITLE
      IF (N.GT.5) WRITE(6,540)
      IF (N.GT.5) KONT=KONT+1
      IF (NG.GT.10) WRITE(6,550)
      IF (NG.GT.10) KONT=KONT+1
      IF (M.GT.7) WRITE(6,560)
      IF (M.GT.7) KONT=KONT+1
      IF (NW.GT.10) WRITE(6,570)
      IF (NW.GT.10) KONT=KONT+1
      IF (NREF.GT.10) WRITE(6,580)
      IF (NREF.GT.10) KONT=KONT+1
      IF (KONT.GT.1) STOP
      DO 10 I=1,NREF
   10 NUMB(I)=I
      DG(1)=GS(1)
      DO 20 I=2,NGG
   20 DG(I)=GS(I)+DG(I-1)
   25 WRITE(6,590) N , NG , SPAN
      WRITE(6,600) E , G
      WRITE(6,610)
      DO 30 I=1,NG
   30 WRITE(6,620) I , GMI(I) , GTI(I)
      WRITE(6,630)
      DO 40 I=1,NGG
   40 WRITE(6,640) I, GS(I),DG(I)
      WRITE(6,650)
      WRITE(6,660) T, EC , GC
      WRITE(6,670) M
```

```
      DO 50 I=1,M
50    WRITE(6,680) I,  W(I) , DLS(I)
      WRITE(6,690) NW
      DO 60 I=1,NW
60    WRITE(6,700) I , DLG(I)
      WRITE(6,710) NREF,(NUMB(I),I=1,NREF)
      WRITE(6,720) (XREF(I),I=1,NREF)
70    CONTINUE
      CALL MOMENT(M,N,W,DLS,SPAN,BM,KZ)
      CALL RMATR(NG,NW,RM,DLG,DG,GS,KZ)
      I1=15
      CALL CONST(I1,NG,EC,GC,T,KT,MLC,C,LAMDA,ALFA,MU,
     *          GMI,GTI,SPAN,GS,G,E,KZ)
      CALL AMATR(I1,NG,DG,GS,AM,KT,MLC,C,LAMDA,ALFA,MU,KZ)
      ICNK=1
      CALL EQN(I1,NG,AM,RM,XX,HARC,KZ,ICNK)
      ICNK=2
      DO 80 I=1,NG
80    CINF(I)=XX(I)
      DO 90 I1=1,N
      CALL CONST(I1,NG,EC,GC,T,KT,MLC,C,LAMDA,ALFA,MU,
     *          GMI,GTI,SPAN,GS,G,E,KZ)
      CALL AMATR(I1,NG,DG,GS,AM,KT,MLC,C,LAMDA,ALFA,MU,KZ)
      CALL EQN(I1,NG,AM,RM,XX,HARC,KZ,ICNK)
90    CONTINUE
      CALL MSDIST(N,NG,SPAN,BM,HARC,ABM,NREF,XREF,M,W,DLS,AS,CINF,KZ)
      CALL FINDEF(N,NG,M,W,DLS,CINF,GMI,E,SPAN,HARC,BM,KZ,NREF,XREF)
510   FORMAT(I1,A)
520   FORMAT(//5X,'SECAN1, ANALYSIS OF BRIDGES BY THE ',/,
     * 5X,'SEMI-CONTINUUM METHOD',//)
530   FORMAT(A)
540   FORMAT(//5X,'HARMONIC NO. LARGER THAN MAXIMUM PERMITTED',//)
550   FORMAT(//5X,'GIRDER NO. LARGER THAN MAXIMUM',//)
560   FORMAT(//5X,'NO. OF LOADS IN ONE LINE LARGER THAN',
     *          /5X,'MAXIMUM PERMITTED',//)
570   FORMAT(//5X,'NO. OF LINES OF LOADS LARGER THAN PERMITTED',//)
580   FORMAT(//5X,'NO. OF REF. SECTION LARGER THAN MAX. PERMITTED',//)
590   FORMAT(//32X,'NUMBER OF HARMONICS   :',I8,/5X,23('*'),
     *       /60X,'NUMBER OF GIRDERS     :',I8,
     *       /60X,'SPAN OF THE BRIDGE    :',F9.1,/)
600   FORMAT(//5X,'GIRDERS INFORMATION   :',/5X,23('*'),
     *        32X,'MODULUS OF ELASTICITY :',E10.4,
     *       //60X,'SHEAR MODULUS         :',E10.4,/)
610   FORMAT(//60X,'GIRDER      MOMENT OF   TORSIONAL ',
     *       /60X,' NO.        INERTIA     INERTIA   ',/)
620   FORMAT( /60X, I3,10X,E10.4,3X,E10.4)
630   FORMAT(/60X,'GIRDER SPACINGS',
     *       //60X,' PANEL       GIRDER    DISTANCE FROM   ',
     *       /60X,' NO.        SPACING    L.H. GIRDER     ',/)
640   FORMAT( /60X, I3, 6X,F8.2,6X,F8.2)
650   FORMAT(1H1//5X,'SLAB INFORMATION      :',/5X,23('*'),
     *        32X,' SLAB       MODULUS OF     SHEAR    ',
     *       /60X,'THICKNESS   ELASTICITY    MODULUS   ',/)
660   FORMAT( /60X,F6.2,6X,2(E10.4,5X),/)
670   FORMAT(//5X,'LOADING INFORMATION   :',32X,'NUMBER OF LOADS IN',
     *                1X,'ONE ',/5X,23('*'),
     *                32X,'LONGITUDINAL LINE IS         :',I6,
     *               //60X,'LOAD    WEIGHT    LOAD DISTANCE FROM',
     *               /60X,' NO.   OF LOAD     L.H. SUPPORT  ',/)
680   FORMAT(/60X,I3, 3X,F9.2 ,5X,F8.2 )
690   FORMAT(//60X,'NUMBER OF LINES OF LOADS IS :',I6,
     *               //60X,'LINE OF    LOAD DISTANCE FROM',
```

```
     *                          /60X,' LOAD        L.H. GIRDER    ',/)
700  FORMAT(/60X,I3,7X,F8.2)
710  FORMAT(//5X,'NO. OF REFERENCE POINTS IS :',I6,
     *          /5X,'REFERENCE POINTS :    ',8(3X,I8,1X))
720  FORMAT(//5X,'DISTANCE FROM L.H. SUPPORT:',8(3X,F8.2,1X))
     STOP
     END
C
C
     SUBROUTINE MSDIST(NT,NG,SPAN,BM,HARC,ABM,NREF,XREF,M,W,DLS,
     *          AS,CINF,KZ)
     DIMENSION BM(5),XREF(10),ABM(10,10),HARC(5,20),NUM(10)
     DIMENSION W(7),DLS(7),AS(10,10),SHR2(10),AMM2(10),CINF(20)
     PI=3.1415926
     DO 10 I=1,NREF
10   NUM(I)=I
     DO 20    I=1,NG
     DO 20    J=1,NREF
     AS(I,J)=0.
     ABM(I,J)=0.
     X=XREF(J)
     CALL MOMSER(X,SHR,AMM,SPAN,M,W,DLS)
     AS(I,J)=SHR*CINF(I)
     ABM(I,J)=AMM*CINF(I)
     SHR2(J)=SHR
     AMM2(J)=AMM
     DO 20    IJ=1,NT
     CONST1=IJ*PI*XREF(J)/SPAN
     CONST2=IJ*PI/SPAN
     ABM(I,J)=ABM(I,J)-BM(IJ)*(CINF(I)-HARC(IJ,I))*SIN(CONST1)
20   AS(I,J)=AS(I,J)-BM(IJ)*(CINF(I)-HARC(IJ,I))*COS(CONST1)*CONST2
     WRITE(6,510) NT
     WRITE(6,520) (NUM(I),I=1,NREF)
     WRITE(6,530) (XREF(I),I=1,NREF)
     WRITE(6,540)
     DO 30    I=1,NG
30   WRITE(6,550) I,(ABM(I,J),J=1,NREF)
     WRITE(6,560)
     DO 40 I=1,NG
40   WRITE(6,570) I,(AS(I,J),J=1,NREF)
     IF(KZ.EQ.1) RETURN
     WRITE(6,580)
     DO 50    J=1,NREF
     WRITE(6,590)XREF(J),SHR2(J),AMM2(J)
50   CONTINUE
510  FORMAT(1H1//5X,'OUTPUT FOR HARMONICS NO. = ',I4)
520  FORMAT(//5X,'MOMENTS IN GIRDERS:',/5X,20('*'),
     *          //5X,'REF. POINT NO.',12X,8(I6,4X))
530  FORMAT(/5X,'DIST. FROM L.H.SUPPORT:',4X,8(F6.1,4X))
540  FORMAT(/5X,'   GIRDER NO. ','                        MOMENTS ',
     *       /5X,80('-'))
550  FORMAT(/5X,I4,21X,8(E10.4,3X))
560  FORMAT(///5X,'SHEARS IN GIRDERS :',
     *          /5X,30('*'),/5X,'GIRDER NO.',27X,'SHEARS'//5X,83('-'))
570  FORMAT(/5X,I5,20X,8(E10.4,3X))
580  FORMAT(//5X,'    X              SHR            AMM',/)
590  FORMAT(5X,3(F14.2,1X))
     RETURN
     END
C
C
     SUBROUTINE MOMSER(X,SHR,AMM,SPAN,M,W,DLS)
```

```
      DIMENSION DLS(7),W(7)
      RL=0.
      DO 10  I=1,M
  10  RL=RL+W(I)*(SPAN-DLS(I))/SPAN
      SHR=RL
      AMM=RL*X
      DO 20 I=1,M
      IF(DLS(I).GE.X) GO TO 30
      SHR=SHR-W(I)
  20  AMM=AMM-W(I)*(X-DLS(I))
  30  CONTINUE
      RETURN
      END
C
C
      SUBROUTINE MOMENT(M,N,W,DLS,SPAN,BM,KZ)
      DIMENSION BM(5),W(7),DLS(7)
      PI=3.1415926
      DO 10 I1=1,N
  10  BM(I1)=0.0
      DO 30 I1=1,N
      DO 20 I3=1,M
      T2=SIN(I1*PI*DLS(I3)/SPAN)
  20  BM(I1)=BM(I1)+(2*SPAN/(I1*I1*PI*PI))*W(I3)*T2
  30  CONTINUE
      IF(KZ.EQ.1) RETURN
      WRITE(6,510)
      WRITE(6,520)
      WRITE(6,530)
      DO 40 I=1,N
  40  WRITE(6,540) I,BM(I)
 510  FORMAT(1H1//)
 520  FORMAT(//10X,'MOMENT COEFFICIENT DUE TO ONE LINE OF WHEELS',
     *        /10X,44('*'),//10X,31('-'),
     *        /10X,'                 ',
     *        /10X,'HARMONICS  ',2X,'MOMENT COEFFICIENT',
     *        /10X,'  NO.          ')
 530  FORMAT(10X,31('-'))
 540  FORMAT(10X,I6,4X,'  ',3X,E10.4)
      RETURN
      END
C
C
      SUBROUTINE RMATR(NG,NW,RM,DLG,DG,GS,KZ)
      DIMENSION RM(20),DLG(10),DG(10),GS(10)
      NG2=2*NG
      DO 10 I1=1,NG2
  10  RM(I1)=0.0
      NGG=NG-1
      DO 40 I1=1,NW
      RM(1)=1.0+RM(1)
      RM(2)=DLG(I1)/DG(NGG)+RM(2)
      DO 20 NP=1,NGG
      NPP=2+NP
      TRM1=(DG(NP)-DLG(I1))/GS(NP)
      IF(TRM1.LT.0.0) TRM1=0.0
      TRM=TRM1*TRM1
      X=TRM
      RM(NPP)=RM(NPP)+X
  20  CONTINUE
      NGB=NG+1
      DO 30 ND=1,NGG
```

```
      NPP=NG+1+ND
      TRM1=(DG(ND)-DLG(I1))/GS(ND)
      TRM=TRM1*TRM1*TRM1
      X=TRM
      IF(X.LT.0.0) X=0.0
      RM(NPP)=RM(NPP)+X
   30 CONTINUE
   40 CONTINUE
      IF (KZ.EQ.1) RETURN
      WRITE(6,510)
  510 FORMAT(//10X,'R VECTOR:',/10X,24('*'),
     *     //10X,35('-'),
     *      /10X,'              ',
     *      /10X,'   ROW       ',5X,'TERM',
     *      /10X,'   NO.       ')
      WRITE(6,520)
  520 FORMAT(10X,35('-'))
      DO 50 I=1,NG2
   50 WRITE(6,530) I,RM(I)
  530 FORMAT(10X,I6,4X,' ',5X,F10.3)
      RETURN
      END
C
C

      SUBROUTINE CONST(I1,NG,EC,GC,T,KT,MLC,C,LAMDA,ALFA,MU,
     *          GMI,GTI,SPAN,GS,G,E,KZ)
      DIMENSION GMI(10),GTI(10),GS(10),MLC(10),C(10),LAMDA(10),
     *          ALFA(10),KT(10),MU(10)
      REAL KT,MLC,LAMDA,MU,EC,E
      PI=3.1415926
      DY=(EC*T**3.)/12.
      DYX=(GC*T**3.)/6.
      DO 20 I2=1,NG
      KT(I2)=((((I1*PI)**4.)*E*GMI(I2))/(SPAN**4.)
      MLC(I2)=((((I1*PI)**2.)*G*GTI(I2))/(SPAN**2.)
      IF(I2.EQ.NG) GO TO 10
      B=GS(I2)
   10 C(I2)=(DYX*(I1*PI)**2.)/(B*SPAN*SPAN)
      LAMDA(I2)=C(I2)/KT(I2)
      ALFA(I2)=(DY*12.)/(KT(I2)*(B**3.))
      MU(I2)=MLC(I2)/(KT(I2)*(B**2.))
      IF (I2.EQ.NG) LAMDA(I2)=0.0
   20 CONTINUE
      IF (KZ.EQ.1) RETURN
      WRITE(6,510) I1
  510 FORMAT(1H1,//5X,'CALCULATED CONSTANTS FOR HARMONICS NO. :',I6,/)
      WRITE(6,520)
  520 FORMAT(/10X,83('-'),
     *    /10X,'  PANEL  ',6('          ')
     *    /10X,'  NO.    ',' ','    K     ',' ','    MLC    ',
     *                     '     C     ',' ',' LAMBDA    ',
     *                     '   ALFA    ',' ',    MU     ')
      WRITE(6,530)
  530 FORMAT(10X,83('-'))
      DO 444 J1=1,NG
      WRITE(6,540)J1,KT(J1),MLC(J1),C(J1,   ..DA(J1),ALFA(J1),MU(J1)
  540 FORMAT(10X,I6,4X,' ',6(E10.4,1X,' ')
  444 CONTINUE
      RETURN
      END
```

```
C
C
      SUBROUTINE AMATR(I1,NG,DG,GS,AM,KT,MLC,C,LAMDA,ALFA,MU,KZ)
      DIMENSION GS(10),MLC(10),C(10),LAMDA(10),
     *    ALFA(10),KT(10),MU(10),DG(10),AM(20,20)
      REAL KT,MLC,LAMDA,MU
      NG2=NG*2
      NGU=NG+1
      NGL=NG-1
      GS(NG)=GS(NGL)
      GS(NGU)=GS(NG)
      DO 10 J1=1,NG2
      DO 10 J2=1,NG2
   10 AM(J1,J2)=0.0
      DO 20 J2=1,NG
   20 AM(1,J2)=1.
      DO 40 J2=1,NG
      IF(J2.EQ.1) GO TO 30
      J2L1=J2-1
      AM(2,J2)=(1./DG(NGL))*((DG(J2L1)+(LAMDA(J2L1)*KT(J2L1)*GS(J2L1)))/
     *KT(J2))-LAMDA(J2)*GS(J2))
      GO TO 40
   30 AM(2,J2)=(1./DG(NGL))*(-LAMDA(J2)*GS(J2))
   40 CONTINUE
      DO 50 J2=1,NG
      NGJ=NG+J2
   50 AM(2,NGJ)=(MU(J2)*GS(J2))/DG(NGL)
C
C     ***  ROWS 2+NP ,WHERE  NP=1,NG-1 ***
C
      DO 120 NP=1,NGL
      NP1=NP+1
      NP2=NP+2
      ARG11=DG(NP)/GS(NP)
      ARG12=ARG11*ARG11
      ARG21=(DG(NP)-DG(1))/GS(NP)
      ARG22=ARG21*ARG21
      AM(NP2,1)=ARG12*(1.+LAMDA(1))-ARG22*LAMDA(1)
      DO 80 NT=2,NP
      NT1=NT-1
      NT2=NT-2
      IF(NT.EQ.2) GO TO 60
      ARGA1=(DG(NP)-DG(NT2))/GS(NP)
      GO TO 70
   60 ARGA1=DG(NP)/GS(NP)
   70 CONTINUE
      ARGA2=ARGA1*ARGA1
      ARGB1=(DG(NP)-DG(NT1))/GS(NP)
      ARGB2=ARGB1*ARGB1
      ARGC1=(DG(NP)-DG(NT))/GS(NP)
      ARGC2=ARGC1*ARGC1
      AM(NP2,NT)=-ARGA2*(LAMDA(NT1)*KT(NT1)/KT(NT))+
     *        ARGB2*(1.+(LAMDA(NT1)*KT(NT1)/KT(NT))+LAMDA(NT))-
     *        (ARGC2*LAMDA(NT))
   80 CONTINUE
   90 CONTINUE
      ARGG1=GS(1)/GS(NP)
      ARGG2=ARGG1*ARGG1
      AM(NP2,NP1)=-(LAMDA(NP)*KT(NP))/KT(NP1)
      AM(NP2,NGU)=-(ALFA(1)/6)*ARGG2-
     *  (2.*MU(1)*(GS(1)/GS(NP))*(DG(NP)/GS(NP)))
```

```
      IF(NP.EQ.1) GO TO 110
      DO 100 NV=2,NP
      NVT=NG+NV
      NVL=NV-1
100   AM(NP2,NVT)=-(2.*MU(NV)*(GS(NV)/GS(NP)))*
     *                ((DG(NP)-DG(NVL))/GS(NP))
110   CONTINUE
      NGP=NG+NP1
120   AM(NP2,NGP)=(ALFA(NP1)/6.)*(GS(NP1)/GS(NP))
      DO 190 ND=1,NGL
      NGUD=NGU+ND
      ARGD1=DG(ND)/GS(ND)
      ARGD3=ARGD1*ARGD1*ARGD1
      ARGE1=(DG(ND)-DG(1))/GS(ND)
      ARGE3=ARGE1*ARGE1*ARGE1
      ARGF1=GS(1)/GS(ND)
      ARGF3=ARGF1*ARGF1*ARGF1
      AM(NGUD,1)=ARGD3*(1.+LAMDA(1))-ARGE3*LAMDA(1)-ARGF3*ALFA(1)/2.
      DO 160 NE=2,ND
      NEL1=NE-1
      NEL2=NE-2
      IF(NE.EQ.2) GO TO 130
      ARGH1=(DG(ND)-DG(NEL2))/GS(ND)
      ARGH3=ARGH1*ARGH1*ARGH1
      GO TO 140
130   ARGH1=DG(ND)/GS(ND)
      ARGH3=ARGH1**3.
140   CONTINUE
      ARGI1=(DG(ND)-DG(NEL1))/GS(ND)
      ARGI3=ARGI1*ARGI1*ARGI1
      ARGJ1=(DG(ND)-DG(NE))/GS(ND)
      ARGJ3=ARGJ1*ARGJ1*ARGJ1
      AM(NGUD,NE)=
     * -ARGH3*(LAMDA(NEL1)*KT(NEL1)/KT(NE))+
     *  ARGI3*((1.+(LAMDA(NEL1)*KT(NEL1)/KT(NE))+LAMDA(NE)))-
     *  (ARGJ3*LAMDA(NE))
160   CONTINUE
      NDU1=ND+1
      ARGM1=GS(NDU1)/GS(ND)
      ARGM3=ARGM1*ARGM1*ARGM1
      ARGN1=GS(1)/GS(ND)
      ARGN2=ARGN1*ARGN1
      ARGO1=DG(ND)/GS(ND)
      ARGO2=ARGO1*ARGO1
      AM(NGUD,NDU1)=-(LAMDA(ND)*KT(ND)/KT(NDU1))+(ALFA(NDU1)/2.)*ARGM3
      AM(NGUD,NGU)=-((ALFA(1)/2.)*ARGN2*(DG(ND)/GS(ND))+
     *              (3.*MU(1)*(GS(1)/GS(ND))*ARGO2))
      IF(ND.EQ.1) GO TO 180
      DO 170 NF=2,ND
      NGNF=NG+NF
      NFL1=NF-1
      NGUD=NG+1+ND
      ARGP1=(DG(ND)-DG(NFL1))/GS(ND)
      ARGP2=ARGP1*ARGP1
      AM(NGUD,NGNF)=-((3.*MU(NF))*(GS(NF)/GS(ND))*ARGP2)
170   CONTINUE
180   CONTINUE
```

```
190 CONTINUE
    IF (KZ.EQ.1) GOTO 210
    WRITE(6,510) I1
    DO 200 JJ=1,NG2
    WRITE(6,520) (AM(JJ,KK),KK=1,NG2)
200 CONTINUE
210 CONTINUE
510 FORMAT(///5X,'CALCULATED AMATRIX FOR HARMONICS NO.:',
    *        I6,/5X,36('*'),///)
520 FORMAT(/5X,10(F8.5,2X))
    RETURN
    END
C
C
    SUBROUTINE EQN(I1,NG,AM,RM,XX,HARC,KZ,ICNK)
    DIMENSION S(20,21),B(20,21),F(20,21),T(20),XX(20),
    * AM(20,20),RM(20),HARC(5,20)
    INTEGER N1
    NG2=NG*2
    NG2U1=NG2+1
    DO 10 I=1,NG2
    XX(I)=0.
    T(I)=0.
    DO 10 J=1,NG2U1
    B(I,J)=0.
    F(I,J)=0.
10  CONTINUE
    DO 20 I=1,NG2
    S(I,NG2U1)=RM(I)
    DO 20 J=1,NG2
    S(I,J)=AM(I,J)
20  CONTINUE
    KI=NG2
    KJ=NG2U1
    DO 30 I=1,KI
    DO 30 J=1,KJ
30  B(I,J)=S(I,J)
    N1=0
40  CONTINUE
    N1=N1+1
    DO 50 I=1,KI
    DO 50 J=1,KJ
50  F(I,J)=B(I,J)
    IF(N2.EQ.KI) GO TO 120
    DO 60 I=N1,KI
60  T(I)=B(I,N1)
    N2=N1+1
    DO 80 I=N1,KI
    DO 80 J=N1,KJ
    IF(ABS(T(I)).EQ.0.0.OR.ABS(F(I,J)).EQ.0.) GO TO 70
    F(I,J)=F(I,J)/T(I)
    GO TO 80
70  F(I,J)=0.0
80  CONTINUE
    DO 90 I=N2,KI
    DO 90 J=N1,KJ
90  F(I,J)=F(N1,J)-F(I,J)
```

```
          DO 110 I=N2,KI
          DO 110 J=1,KJ
 110  B(I,J)=F(I,J)
          GO TO 40
 120  CONTINUE
          MB=KI
          NB=KJ
          DO 130 J=1,NB
 130  B(1,J)=S(1,J)
          INB=NB-1
          IF(ABS(B(MB,NB)).EQ.0.0.OR.ABS(B(MB,MB)).EQ.0.0) GO TO 140
          XX(MB)=B(MB,NB)/B(MB,MB)
          GO TO 150
 140  XX(MB)=0.0
 150  CONTINUE
          N3=1
 160  MB=KI
          MB=MB-N3
          KK=NB-N3
          TEXP=0.
          DO 170 J=KK,INB
 170  TEXP=TEXP+XX(J)*B(MB,J)
          BWT=ABS(B(MB,NB)-TEXP)
          IF(BWT.EQ.0.0.OR.ABS(B(MB,MB)).EQ.0.0) GO TO 180
          XX(MB)=(B(MB,NB)-TEXP)/B(MB,MB)
          GO TO 190
 180  XX(MB)=0.0
 190  CONTINUE
          N3=N3+1
          IF(N3.EQ.KI) GO TO 200
          GO TO 160
 200  CONTINUE
          IF(ICNK.EQ.1) GO TO 220
          DO 210 I=1,KI
 210  HARC(I1,I)=XX(I)
 220  CONTINUE
          IF(KZ.EQ.1) GO TO 250
          WRITE(6,510)
          DO 240 I=1,KI
          WRITE(6,520) I,XX(I)
 240  CONTINUE
 250  CONTINUE
          DO 260 I=1,20
          T(I)=0.
          DO 260 J=1,21
          B(I,J)=0.
          F(I,J)=0.
          S(I,J)=0.
 260  CONTINUE
          N2=0
 510  FORMAT(//,15X,'CORRELATION COEFFICIENTS ',/15X,26('*'),//)
 520  FORMAT(15X,'B',I2,' = ',F14.5)
          RETURN
          END
C
C

          SUBROUTINE DEFLEC(GMI,E,X1,I1,SPAN,W,DLS,WT,M)

          DIMENSION GMI(10),W(7),DLS(7)
```

```
       WT=0.0
       DO 30 I=1,M
       X=X1
       IF(X .GT. DLS(I)) GOTO 10
       A=DLS(I)
       B=SPAN-A
       GOTO 20
 10    B=DLS(I)
       A=SPAN-B
       X=SPAN-X1
 20    CONTINUE
       WT=WT+W(I)*B*(A*SPAN*X+A*B*X-X*X*X)/(6*E*GMI(I1)*SPAN)
 30    CONTINUE
       RETURN
       END
C
C
       SUBROUTINE FINDEF(NN,NG,M,W,DLS,CINF,GMI,E,
      *SPAN,HARC,BM,KZ,NREF,XREF)
       DIMENSION W(7),DLS(7),CINF(20),
      *          GMI(10),HARC(5,20),BM(5),WF(10,10),
      *          NUM(10),XREF(10)
       DO 10 I=1,10
       DO 10 J=1,10
 10    WF(I,J)=0.0
       PI=3.141592654
       DO 30 I=1,NREF
       DO 30 J=1,NG
       CIN=CINF(J)
       X=XREF(I)
       CALL DEFLEC(GMI,E,X,J,SPAN,W,DLS,WT,M)
       YY=WT*CIN
       DO 20 IJ=1,NN
       CONST1=IJ*PI*X/SPAN
       YY=YY-BM(IJ)*(CIN-HARC(IJ,J))*SIN(CONST1)*SPAN*SPAN/(E*GMI(J)*
      *PI*PI*IJ*IJ)
 20    CONTINUE
 30    WF(J,I)=YY
       DO 70 I=1,NREF
 70    NUM(I)=I
       WRITE(6,510) (NUM(I),I=1,NREF)
       WRITE(6,520) (XREF(I),I=1,NREF)
       WRITE(6,530)
       DO 80 I=1,NG
 80    WRITE(6,540) I,(WF(I,J),J=1,NREF)
 510   FORMAT(//5X,'DEFLECTIONS IN GIRDERS:',
      */5X,20('*'),//5X,'REF. POINT NO.',12X,8(I6,4X))
 520   FORMAT(/,5X,'DIST. FROM L.H. SUPPORT:',4X,8(F6.1,4X))
 530   FORMAT(/,5X,'    GIRDER NO.',20X,'DEFLECTIONS',/,5X,80('-'))
 540   FORMAT(/,5X,I4,21X,8(E10.4,3X))
       RETURN
       END
```

VII

Listings of SECAN2

This program analyzes slab-on-girder bridges with right, simply supported ends and random intermediate supports by the semicontinuum method.

```
C           SECAN2
C
C      THIS PROGRAM ANALYSES SLAB-ON-GIRDER BRIDGES
C      WITH INTERMEDIATE SUPPORTS, BY THE
C      SEMICONTINUUM METHOD


       DIMENSION GS(10),GMI(10),GTI(10),
      *DLS(7),DG(10),XX(20),DLG(10),W(7)
       DIMENSION AM(20,20),BM(5),RM(20),
      *KT(10),MLC(10),C(10),LAMDA(10),ALFA(10),MU(10),NUMB(10),
      *DELTA(10),EF(10),WB(10),WU(10,10),ARE(10),KGIR(10),XCOL(10),

      *DLGI(10),BI(10,11),DI(10),HARC1(5,20),W1(7),DLSI(7),
      *BM1(5),FF(10),DLS1(7)
       DIMENSION XREF(10),HARC(5,20),ABM(10,10),AS(10,10),CINF(20,11)

       REAL KT,MLC,MU,LAMDA
       CHARACTER*52 TITLE
       READ(5,510) KZ,TITLE
C
C      1 MAKE KZ=2 IF INTERMEDIATE RESULTS ARE REQUIRED,
C      2 OTHERWISE MAKE IT EQUAL TO 1
C
       READ(5,*) N,NG,SPAN,E,G,NCOL
       NGG=NG-1
       READ(5,*) (GS(I),I=1,NGG)
       READ(5,*) (GMI(I),I=1,NG)
       READ(5,*) (GTI(I),I=1,NG)
       READ(5,*) T,EC,GC
       READ(5,*) M
       READ(5,*) (W(I),I=1,M)
       READ(5,*) (DLS(J),J=1,M)
       READ(5,*) NW
       READ(5,*) (DLG(I),I=1,NW)
       READ(5,*) NREF
       READ(5,*) (XREF(I),I=1,NREF)
       IF (NCOL .EQ. 0) GOTO 10
       READ(5,*) (DELTA(I),I=1,NCOL)
       READ(5,*) (FF(I),I=1,NCOL)
       READ(5,*) (KGIR(I),I=1,NCOL)
       READ(5,*) (XCOL(I),I=1,NCOL)
       GO TO 15
   10  WRITE(6,505)
       STOP
   15  CONTINUE
       WRITE(6,520)
       WRITE(6,530) TITLE
       IF (N.GT.5) WRITE(6,540)
       IF (N.GT.5) KONT=KONT+1
       IF (NG.GT.10) WRITE(6,550)
       IF (NG.GT.10) KONT=KONT+1
       IF (M.GT.7) WRITE(6,560)
       IF (M.GT.7) KONT=KONT+1
       IF (NW.GT.10) WRITE(6,570)
       IF (NW.GT.10) KONT=KONT+1
       IF (NREF.GT.10) WRITE(6,580)
       IF (NREF.GT.10) KONT=KONT+1
```

```
      IF (NCOL .GT. 10) WRITE(6,590)
      IF (NCOL .GT. 10) KONT=KONT+1
      IF (KONT.GT.1) STOP
      DO 20 I=1,NREF
   20 NUMB(I)=I
      DG(1)=GS(1)
      DO 30 I=2,NGG

   30 DG(I)=GS(I)+DG(I-1)
      WRITE(6,600) N , NG , SPAN
      WRITE(6,610) E , G ,NCOL
      WRITE(6,620)
      DO 40 I=1,NG

   40 WRITE(6,630) I , GMI(I) , GTI(I)
      WRITE(6,640)
      DO 50 I=1,NGG

   50 WRITE(6,650) I, GS(I),DG(I)
      WRITE(6,660)
      WRITE(6,670) T, EC , GC
      WRITE(6,680) M
      DO 60 I=1,M

   60 WRITE(6,690) I,  W(I) , DLS(I)
      WRITE(6,800) NW
      DO 70 I=1,NW
   70 WRITE(6,700) I , DLG(I)
      WRITE(6,710) NREF,(NUMB(I),I=1,NREF)
      WRITE(6,720) (XREF(I),I=1,NREF)
      WRITE (6,730) (DELTA(I),I=1,NCOL)
      WRITE (6,770)
      WRITE (6,740) (FF(I),I=1,NCOL)
      WRITE (6,750) (KGIR(I),I=1,NCOL)
      WRITE (6,760) (XCOL(I),I=1,NCOL)
      WRITE (6,780)
   80 CONTINUE
      CALL MOMENT(M,N,W,DLS,SPAN,BM,KZ)
      CALL RMATR(NG,NW,RM,DLG,DG,GS,KZ)
      I1=15
      CALL CONST(I1,NG,EC,GC,T,KT,MLC,C,LAMDA,ALFA,MU,
     *          GMI,GTI,SPAN,GS,G,E,KZ)
      CALL AMATR(I1,NG,DG,GS,AM,KT,MLC,C,LAMDA,ALFA,MU,KZ)
      ICNK=1
      CALL EQN(I1,NG,AM,RM,XX,HARC,KZ,ICNK)
      ICNK=2
      DO 90 I=1,NG
      CINF(I,1)=XX(I)
   90 CONTINUE
C
      DO 100 I1=1,N

      CALL CONST(I1,NG,EC,GC,T,KT,MLC,C,LAMDA,ALFA,MU,
     *          GMI,GTI,SPAN,GS,G,E,KZ)
      CALL AMATR(I1,NG,DG,GS,AM,KT,MLC,C,LAMDA,ALFA,MU,KZ)
      CALL EQN(I1,NG,AM,RM,XX,HARC,KZ,ICNK)
  100 CONTINUE
C
```

```
C OBTAIN DEFLECTIONS AT COLUMN LOCATIONS
C DUE TO APPLIED LOADS
      LCONT=1
      CALL WDIST(N,NCOL,KGIR,XCOL,WB,M,W,DLS,CINF,GMI,E,SPAN,HARC,
     *BM,LCONT,WU,I3,KZ)
C ARRAY WB CONTAINS DEFLECTIONS AT COLUMN LOCATIONS
C OBTAIN DEFLECTIONS AT COLUMN LOCATIONS DUE TO UNIT LOAD AT EACH
C COLUMN
C ARRAY WU(I,J) CONTAINS THE DEFLECTION. THE FIRST SUBSCRIPT REFERS
C TO THE COLUMN NUMBER AT WHICH THE DEFLECTIONS ARE SOUGHT AND THE
C SECOND TO THE COLUMN NUMBER AT WHICH THE LOAD IS APPLIED.
C
      DO 110 I1=1,10
  110 DLGI(I1)=0.0
C
      NWI=1
C
      DO 160 I3=1,NCOL
      I4 = KGIR(I3)
      IF (I4 .LT. NG) DLGI(1)=DG(I4)-GS(I4)
      IF (I4 .EQ. NG)DLGI(1)=DG(I4-1)
      CALL RMATR(NG,NWI,RM,DLGI,DG,GS,KZ)
      NPLUS=N+1
      DO 150 I7=1,NPLUS
      IF (I7 .GT. 1) GOTO 120
      I1=15
      ICNK=1
      GOTO 130
  120 I1=I7-1
      ICNK=2
  130 CONTINUE
      CALL CONST(I1,NG,EC,GC,T,KT,MLC,C,LAMDA,ALFA,MU,GMI,GTI,SPAN,
     *GS,G,E,KZ)
      CALL AMATR(I1,NG,DG,GS,AM,KT,MLC,C,LAMDA,ALFA,MU,KZ)
      CALL EQN(I1,NG,AM,RM,XX,HARC1,KZ,ICNK)
      IF (I7 .GT. 1) GOTO 150
      I9=I3+1
      DO 140 I8=1,NG
  140 CINF(I8,I9)=XX(I8)
  150 CONTINUE
      M1=1
      W1(1)=1.0
      DLS1(1)=XCOL(I3)
      CALL MOMENT(M1,N,W1,DLS1,SPAN,BM1,KZ)
      LCONT=2
      CALL WDIST(N,NCOL,KGIR,XCOL,WB,M1,W1,DLS1,CINF,GMI,E,SPAN
     *,HARC1,BM1,LCONT,WU,I3,KZ)
  160 CONTINUE
C
      CALL BEE(FF,WU,BI,NCOL)
      CALL DEE(WB,DELTA,DI,NCOL)
      CALL SOLVE(BI,NCOL,DI)
      DO 170 MM=1,NCOL
  170 ARE(MM)=DI(MM)
      WRITE (6,790) (ARE(I3),I3=1,NCOL)
C MSDIST IS CALLED TO CALCULATE MOMENT AND SHEARS FOR THE CASE
C WITHOUT INTERMEDIATE SUPPORTS
      CALL MSDIST(N,NG,SPAN,BM,HARC,ABM,NREF,XREF,M,W,DLS,AS,CINF,KZ)
C
```

```
      DO 190 I3=1,NCOL
      I4 = KGIR(I3)
      IF (I4 .LT. NG) DLGI(1)=DG(I4)-GS(I4)
      IF (I4 .EQ. NG)DLGI(1)=DG(I4-1)
      CALL RMATR(NG,NWI,RM,DLGI,DG,GS,KZ)
      DO 180 I1=1,N
      CALL CONST(I1,NG,EC,GC,T,KT,MLC,C,LAMDA,ALFA,MU,GMI,GTI,SPAN,GS,
     *G,E,KZ)
      CALL AMATR(I1,NG,DG,GS,AM,KT,MLC,C,LAMDA,ALFA,MU,KZ)
      CALL EQN(I1,NG,AM,RM,XX,HARC1,KZ,ICNK)
  180 CONTINUE
      M1=1
      W1(1)=-ARE(I3)
      DLSI(1)=XCOL(I3)
      CALL MOMENT(M1,N,W1,DLSI,SPAN,BM1,KZ)
      CALL MSD2(I3,N,NG,SPAN,BM1,HARC1,ABM,NREF,XREF,M1,W1,DLSI,AS,
     *CINF,KZ,NCOL)
      CALL FINDEF(I3,N,NCOL,NG,KGIR,XCOL,M,W,ARE,DLS,CINF,GMI,E,
     *SPAN,HARC,HARC1,BM,KZ,NREF,XREF)
  190 CONTINUE
C
C          FORMAT STATEMENTS
C
  505 FORMAT(/,' SECAN2 CANNOT ANALYSE BRIDGES WITHOUT INTERMEDIATE ',
     * /,'SUPPORTS----FOR THIS BRIDGE USE SECAN1 ')
  510 FORMAT(I1,A)
  520 FORMAT(//5X,'SECAN2,ANALYSIS OF BRIDGES WITH RANDOM INTERMEDIATE',
     */,' SUPPORTS BY THE SEMI-CONTINUUM METHOD',/)
  530 FORMAT(A)
  540 FORMAT(//5X,'HARMONIC NO. LARGER THAN MAXIMUM PERMITTED',//)
  550 FORMAT(//5X,'GIRDER NO. LARGER THAN MAXIMUM',//)
  560 FORMAT(//5X,'NO. OF LOADS IN ONE LINE LARGER THAN',
     *        /5X,'MAXIMUM PERMITTED',//)
  570 FORMAT(//5X,'NO. OF LINES OF LOADS LARGER THAN PERMITTED',//)
  580 FORMAT(//5X,'NO. OF REF. SECTIONS LARGER THAN MAX. PERMITTED',//)
  590 FORMAT(//,5X,'NO. OF COLUMNS IS GREATER THAN PERMITTED',//)
  600 FORMAT(//32X,'NUMBER OF HARMONICS   :',I8,/,5X,23('*'),
     *        /60X,'NUMBER OF GIRDERS     :',I8,
     *        /60X,'SPAN OF THE BRIDGE    :',F9.1,/)
  610 FORMAT(//5X,'GIRDERS INFORMATION   :',/5X,23('*'),
     *       32X,'MODULUS OF ELASTICITY :',E10.4,
     *       //60X,'SHEAR MODULUS         :',E10.4,/
     * /60X,'NO. OF COLUMNS:',I5,/)
  620 FORMAT(//60X,'GIRDER       MOMENT OF    TORSIONAL ',
     *        /60X,' NO.         INERTIA      INERTIA   ',/)
  630 FORMAT( /60X, I3,10X,E10.4,3X,E10.4)
  640 FORMAT(/60X,'GIRDER SPACINGS',
     *       //60X,' PANEL       GIRDER     DISTANCE FROM   ',
     *       /60X,' NO.         SPACING    L.H. GIRDER     ',/)
  650 FORMAT( /60X, I3, 6X,F8.2,6X,F8.2)
  660 FORMAT(1H1//5X,'SLAB INFORMATION      :',/5X,23('*'),
     *       32X,' SLAB       MODULUS OF      SHEAR    ',
     *       /60X,'THICKNESS   ELASTICITY     MODULUS  ',/)
  670 FORMAT( /60X,F6.2,6X,2(E10.4,5X),/)
  680 FORMAT(//5X,'LOADING INFORMATION    :',32X,'NUMBER OF LOAD IN',
     *               1X,'ONE ',/5X,23('*'),
     *               32X,'LONGITUDINAL LINE IS          :',I6,
     *               //60X,'LOAD   WEIGHT   LOAD DISTANCE FROM',
     *               /60X,' NO.   OF LOAD   L.H. SUPPORT ',/)
  690 FORMAT(/60X,I3, 3X,F9.2 ,5X,F8.2 )
```

```
700   FORMAT(/60X,I3,7X,F8.2)
710   FORMAT(//5X,'NO. OF REFERENCE POINTS IS :',I6,
      *           /5X,'REFERENCE POINTS  :   ',8(3X,I8,1X))
720   FORMAT(//5X,'DISTANCE FROM LH. SUPPORT :',8(3X,F8.2,1X))
730   FORMAT(//,5X,'COLUMN DETAILS',/,5X,'**************',
      *//,5X,'PRESCRIBED',4X,8(3X,F8.2,1X))
740   FORMAT(//,5X,'COL.FLEXIBILITY',8(3X,F8.2,1X))
750   FORMAT(//,5X,'GIRDER NUMBER',2X,8(3X,I8,1X))
760   FORMAT(//,5X,'DISTANCE FROM',2X,8(3X,F8.1,1X))
770   FORMAT(/,5X,'DEFLECTIONS')
780   FORMAT(/,5X,'L. H. SUPPORT')
790   FORMAT (//5X,'COLUMN REACTIONS',10F10.3)
800   FORMAT(//60X,'NUMBER OF LINES OF LOADS IS :',I6,
      *                 //60X,'LINE OF    LOAD DISTANCE FROM',
      *                 /60X,' LOAD         L.H. GIRDER   ',/)
C
      STOP
      END
C
C
      SUBROUTINE MSDIST(NT,NG,SPAN,BM,HARC,ABM,NREF,XREF,M,W,DLS,
      *            AS,CINF,KZ)
      DIMENSION BM(5),XREF(10),ABM(10,10),HARC(5,20),NUM(10)
      DIMENSION W(7),DLS(7),AS(10,10),SHR2(10),AMM2(10),CINF(20,11)
      PI=3.141592654
      DO 10 I=1,NREF
   10 NUM(I)=I
      DO 20 I=1,NG
      DO 20 J=1,NREF
      AS(I,J)=0.
      ABM(I,J)=0.
      X=XREF(J)
      CALL MOMSER(X,SHR,AMM,SPAN,M,W,DLS)
      AS(I,J)=SHR*CINF(I,1)
      ABM(I,J)=AMM*CINF(I,1)
      SHR2(J)=SHR
      AMM2(J)=AMM
      DO 20 IJ=1,NT
      CONST1=IJ*PI*XREF(J)/SPAN
      CONST2=IJ*PI/SPAN
      ABM(I,J)=ABM(I,J)-BM(IJ)*(CINF(I,1)-HARC(IJ,I))*SIN(CONST1)
   20 AS(I,J)=AS(I,J)-BM(IJ)*(CINF(I,1)-HARC(IJ,I))*COS(CONST1)*CONST2
      IF(KZ.EQ.1) RETURN
      WRITE(6,520) NT
      WRITE(6,530) (NUM(I),I=1,NREF)
      WRITE(6,540) (XREF(I),I=1,NREF)
      WRITE(6,550)
      DO 30 I=1,NG
   30 WRITE(6,560) I,(ABM(I,J),J=1,NREF)
      WRITE(6,570)
      DO 40 I=1,NG
   40 WRITE(6,580) I,(AS(I,J),J=1,NREF)
  520 FORMAT(1H1//5X,'OUTPUT FOR HARMONICS NO. = ',I4)
  530 FORMAT(//5X,'MOMENTS IN GIRDERS:',/5X,20('*'),
      *         //5X,'REF. POINT NO.',12X,8(I6,4X))
  540 FORMAT(/5X,'DIST. FROM L.H.SUPPORT:',4X,8(F6.1,4X))
  550 FORMAT(/5X,'   GIRDER NO. ','                        MOMENTS ',
      *        /5X,80('-'))
  560 FORMAT(/5X,I4,21X,8(E10.4,3X))
  570 FORMAT(///5X,' SHEARS IN GIRDERS :',
```

```
   *              /5X,30('*'),/5X,'GIRDER NO.',27X,'SHEARS'//5X,83('-'))
 580 FORMAT(/5X,I5,20X,8(E10.4,3X))
     RETURN
     END
C
C
     SUBROUTINE MOMSER(X,SHR,AMM,SPAN,M,W,DLS)
     DIMENSION DLS(7),W(7)
     RL=0.
     DO 10 I=1,M
  10 RL=RL+W(I)*(SPAN-DLS(I))/SPAN
     SHR=RL
     AMM=RL*X
     DO   20 I=1,M
     IF(DLS(I).GE.X) GO TO 30
     SHR=SHR-W(I)
  20 AMM=AMM-W(I)*(X-DLS(I))
  30 CONTINUE
     RETURN
     END
C
C
     SUBROUTINE MOMENT(M,N,W,DLS,SPAN,BM,KZ)
     DIMENSION BM(5),W(7),DLS(7)
     PI=3.141592654
     DO 10 I1=1,N
  10 BM(I1)=0.0
     DO 30 I1=1,N
     DO 20 I3=1,M
     T2=SIN(I1*PI*DLS(I3)/SPAN)
  20 BM(I1)=BM(I1)+(2*SPAN/(I1*I1*PI*PI))*W(I3)*T2
  30 CONTINUE
     IF(KZ.EQ.1) RETURN
     WRITE(6,510)
     WRITE(6,520)
     WRITE(6,530)
     DO 40 I=1,N
  40 WRITE(6,540) I,BM(I)
 510 FORMAT(1H1//)
 520 FORMAT(//10X,'MOMENT COEFFICIENTS DUE TO ONE LINE OF WHEELS',
   *              /10X,44('*'),//10X,31('-'),
   *              /10X,'            |',
   *              /10X,'HARMONICS |',2X,'MOMENT COEFFICIENT',
   *              /10X,' NO.     |')
 530 FORMAT(10X,31('-'))
 540 FORMAT(10X,I6,4X,'|',3X,E10.4)
     RETURN
     END
C
C
     SUBROUTINE RMATR(NG,NW,RM,DLG,DG,GS,KZ)
     DIMENSION RM(20),DLG(10),DG(10),GS(10)
     NG2=2*NG
     DO 10 I1=1,NG2
  10 RM(I1)=0.0
     NGG=NG-1
     DO 40 I1=1,NW
     RM(1)=1.0+RM(1)
     RM(2)=DLG(I1)/DG(NGG)+RM(2)
```

```
      DO 20 NP=1,NGG
      NPP=2+NP
      TRM1=(DG(NP)-DLG(I1))/GS(NP)
      IF(TRM1.LT.0.0) TRM1=0.0
      TRM=TRM1*TRM1
      X=TRM
      RM(NPP)=RM(NPP)+X
   20 CONTINUE
      DO 30 ND=1,NGG
      NPP=NG+1+ND
      TRM1=(DG(ND)-DLG(I1))/GS(ND)
      TRM=TRM1*TRM1*TRM1
      X=TRM
      IF(X.LT.0.0) X=0.0
      RM(NPP)=RM(NPP)+X
   30 CONTINUE
   40 CONTINUE
      IF (KZ.EQ.1) RETURN
      WRITE(6,510)
      WRITE(6,520)
      DO 50 I=1,NG2
   50 WRITE(6,530) I,RM(I)
  510 FORMAT(//10X,'CALCULATED R VECTOR :',/10X,24('*'),
     *     //10X,35('-'),
     *     /10X,'                    |',
     *     /10X,'   ROW              |',5X,'TERM',
     *     /10X,'   NO.              |')
  520 FORMAT(10X,35('-'))
  530 FORMAT(10X,I6,4X,'|',5X,F10.3)
      RETURN
      END
C
C
      SUBROUTINE CONST(I1,NG,EC,GC,T,KT,MLC,C,LAMDA,ALFA,MU,
     *          GMI,GTI,SPAN,GS,G,E,KZ)
      DIMENSION GMI(10),GTI(10),GS(10),MLC(10),C(10),LAMDA(10),
     *          ALFA(10),KT(10),MU(10)
      REAL KT,MLC,LAMDA,MU,EC,E
      PI=3.141592654
      DY=(EC*T**3.)/12.
      DYX=(GC*T**3.)/6.
      DO 20 I2=1,NG
      KT(I2)=((((I1*PI)**4.)*E*GMI(I2))/(SPAN**4.)
      MLC(I2)=((((I1*PI)**2.)*G*GTI(I2))/(SPAN**2.)
      IF(I2.EQ.NG) GO TO 10
      B=GS(I2)
   10 C(I2)=(DYX*(I1*PI)**2.)/(B*SPAN*SPAN)
      LAMDA(I2)=C(I2)/KT(I2)
      ALFA(I2)=(DY*12.)/(KT(I2)*(B**3.))
      MU(I2)=MLC(I2)/(KT(I2)*(B**2.))
      IF (I2.EQ.NG) LAMDA(I2)=0.0
   20 CONTINUE
      IF (KZ.EQ.1) RETURN
      WRITE(6,510) I1
      WRITE(6,520)
      WRITE(6,530)
      DO 30 J1=1,NG
      WRITE(6,540)J1,KT(J1),MLC(J1),C(J1),LAMDA(J1),ALFA(J1),MU(J1)
   30 CONTINUE
  510 FORMAT(1H1,//5X,'CALCULATED CONSTANTS FOR HARMONICS NO. :',I6,/)
```

```
  520 FORMAT(/10X,83('-'),
     *      /10X,'   PANEL  |',6('                      |')
     *      /10X,'    NO.   |',  '      K      |',' MLC     |',
     *                         '      C      |',' LAMBDA  |',
     *                         ' ALFA        |',' MU      |')
  530 FORMAT(10X,83('-'))
  540 FORMAT(10X,I6,4X,'|',6(E10.4,1X,'|'))
      RETURN
      END
C
C
      SUBROUTINE AMATR(I1,NG,DG,GS,AM,KT,MLC,C,LAMDA,ALFA,MU,KZ)
      DIMENSION GS(10),MLC(10),C(10),LAMDA(10),
     *    ALFA(10),KT(10),MU(10),DG(10),AM(20,20)
      REAL KT,MLC,LAMDA,MU
      NG2=NG*2
      NGU=NG+1
      NGL=NG-1
      GS(NG)=GS(NGL)
      GS(NGU)=GS(NG)
      DO 10 J1=1,NG2
      DO 10 J2=1,NG2
   10 AM(J1,J2)=0.0
      DO 20 J2=1,NG
   20 AM(1,J2)=1.
      DO 40 J2=1,NG
      IF(J2.EQ.1) GO TO 30
      J2L1=J2-1
      AM(2,J2)=(1./DG(NGL))*((DG(J2L1)+(LAMDA(J2L1)*KT(J2L1)*GS(J2L1))/
     *KT(J2))-LAMDA(J2)*GS(J2))
      GO TO 40
   30 AM(2,J2)=(1./DG(NGL))*(-LAMDA(J2)*GS(J2))
   40 CONTINUE
      DO 50 J2=1,NG
      NGJ=NG+J2
   50 AM(2,NGJ)=(MU(J2)*GS(J2))/DG(NGL)
C
C     *** ROWS 2+NP ,WHERE  NP=1,NG-1 ***
C
      DO 110 NP=1,NGL
      NP1=NP+1
      NP2=NP+2
      ARG11=DG(NP)/GS(NP)
      ARG12=ARG11*ARG11
      ARG21=(DG(NP)-DG(1))/GS(NP)
      ARG22=ARG21*ARG21
      AM(NP2,1)=ARG12*(1.+LAMDA(1))-ARG22*LAMDA(1)
      DO 80 NT=2,NP
      NT1=NT-1
      NT2=NT-2
      IF(NT.EQ.2) GO TO 60
      ARGA1=(DG(NP)-DG(NT2))/GS(NP)
      GO TO 70
   60 ARGA1=DG(NP)/GS(NP)
   70 CONTINUE
      ARGA2=ARGA1*ARGA1
      ARGB1=(DG(NP)-DG(NT1))/GS(NP)
      ARGB2=ARGB1*ARGB1
      ARGC1=(DG(NP)-DG(NT))/GS(NP)
      ARGC2=ARGC1*ARGC1
```

```
      AM(NP2,NT)=-ARGA2*(LAMDA(NT1)*KT(NT1)/KT(NT))+
     *       ARGB2*(1.+(LAMDA(NT1)*KT(NT1)/KT(NT))+LAMDA(NT))-
     *       (ARGC2*LAMDA(NT))
  80 CONTINUE
      ARGG1=GS(1)/GS(NP)
      ARGG2=ARGG1*ARGG1
      AM(NP2,NP1)=-(LAMDA(NP)*KT(NP))/KT(NP1)
      AM(NP2,NGU)=-(ALFA(1)/6)*ARGG2-
     *   (2.*MU(1)*(GS(1)/GS(NP))*(DG(NP)/GS(NP)))
      IF(NP.EQ.1) GO TO 100
      DO 90 NV=2,NP
      NVT=NG+NV
      NVL=NV-1
  90 AM(NP2,NVT)=-(2.*MU(NV)*(GS(NV)/GS(NP)))*
     *                ((DG(NP)-DG(NVL))/GS(NP))
 100 CONTINUE
      NGP=NG+NP1
 110 AM(NP2,NGP)=(ALFA(NP1)/6.)*(GS(NP1)/GS(NP))
      DO 160 ND=1,NGL
      NGUD=NGU+ND
      ARGD1=DG(ND)/GS(ND)
      ARGD3=ARGD1*ARGD1*ARGD1
      ARGE1=(DG(ND)-DG(1))/GS(ND)
      ARGE3=ARGE1*ARGE1*ARGE1
      ARGF1=GS(1)/GS(ND)
      ARGF3=ARGF1*ARGF1*ARGF1
      AM(NGUD,1)=ARGD3*(1.+LAMDA(1))-ARGE3*LAMDA(1)-ARGF3*ALFA(1)/2.
      DO 135 NE=2,ND
      NEL1=NE-1
      NEL2=NE-2
      IF(NE.EQ.2) GO TO 120
      ARGH1=(DG(ND)-DG(NEL2))/GS(ND)
      ARGH3=ARGH1*ARGH1*ARGH1
      GO TO 130
 120 ARGH1=DG(ND)/GS(ND)
      ARGH3=ARGH1**3.
 130 CONTINUE
      ARGI1=(DG(ND)-DG(NEL1))/GS(ND)
      ARGI3=ARGI1*ARGI1*ARGI1
      ARGJ1=(DG(ND)-DG(NE))/GS(ND)
      ARGJ3=ARGJ1*ARGJ1*ARGJ1
      AM(NGUD,NE)=
     * -ARGH3*(LAMDA(NEL1)*KT(NEL1)/KT(NE))+
     *  ARGI3*((1.+(LAMDA(NEL1)*KT(NEL1)/KT(NE))+LAMDA(NE)))-
     *  (ARGJ3*LAMDA(NE))
 135 CONTINUE
      NDU1=ND+1
      ARGM1=GS(NDU1)/GS(ND)
      ARGM3=ARGM1*ARGM1*ARGM1
      ARGN1=GS(1)/GS(ND)
      ARGN2=ARGN1*ARGN1
      ARGO1=DG(ND)/GS(ND)
      ARGO2=ARGO1*ARGO1
      AM(NGUD,NDU1)=-(LAMDA(ND)*KT(ND)/KT(NDU1))+(ALFA(NDU1)/2.)*ARGM3
      AM(NGUD,NGU)=-((ALFA(1)/2.)*ARGN2*(DG(ND)/GS(ND))+
     *                (3.*MU(1)*(GS(1)/GS(ND))*ARGO2))
      IF(ND.EQ.1) GO TO 150
      DO 140 NF=2,ND
      NGNF=NG+NF
      NFL1=NF-1
```

```
      NGUD=NG+1+ND
      ARGP1=(DG(ND)-DG(NFL1))/GS(ND)
      ARGP2=ARGP1*ARGP1
      AM(NGUD,NGNF)=-((3.*MU(NF))*(GS(NF)/GS(ND))*ARGP2)
  140 CONTINUE
  150 CONTINUE
  160 CONTINUE
      IF (KZ.EQ.1) GOTO 180
      WRITE(6,510) I1
      DO 170 JJ=1,NG2
      WRITE(6,520) (AM(JJ,KK),KK=1,NG2)
  170 CONTINUE
  180 CONTINUE
  510 FORMAT(///5X,'CALCULATED AMATRIX FOR HARMONICS NO.:',
     *        I6,/5X,36('*'),///)
  520 FORMAT(/5X,10(F8.5,2X))
      RETURN
      END
C
C
      SUBROUTINE MSD2(I3,N,NG,SPAN,BM,HARC1,ABM,NREF,XREF,
     *M1,W1,DLS1,AS,CINF,KZ,NCOL)
      DIMENSION W1(7),DLS1(7),AS(10,10),BM(5),XREF(10),ABM(10,10),
     *HARC1(5,20),CINF(20,11),NUM(10),SHR2(10),AMM2(10)
      I3P=I3+1
      PI=3.141592654
      DO 10 I=1,NREF

   10 NUM(I)=I
      DO 20 I=1,NG
      DO 20 J=1,NREF
      X=XREF(J)
      CALL MOMSER(X,SHR,AMM,SPAN,M1,W1,DLS1)
      CS=SHR*CINF(I,I3P)
      CM=AMM*CINF(I,I3P)
      SHR2(J)=SHR
      AMM2(J)=AMM
      DO 20 IJ=1,N
      CONST1=IJ*PI*XREF(J)/SPAN
      CONST2=IJ*PI/SPAN
      CM=CM-BM(IJ)*(CINF(I,I3P)-HARC1(IJ,I))*SIN(CONST1)
      CS=CS-BM(IJ)*(CINF(I,I3P)-HARC1(IJ,I))*COS(CONST1)*CONST2
      IF (IJ.EQ.N) ABM(I,J)=ABM(I,J)+CM
      IF (IJ.EQ.N) AS(I,J)=AS(I,J)+CS
   20 CONTINUE
      IF(I3 .NE. NCOL) GOTO 30
      WRITE(6,510)
      GOTO 40
   30 IF (KZ .EQ. 1) RETURN
      WRITE(6,600) N
   40 CONTINUE
      WRITE(6,520) (NUM(I),I=1,NREF)
      WRITE(6,530) (XREF(I),I=1,NREF)
      WRITE(6,540)
      DO 50 I=1,NG
   50 WRITE(6,550) I,(ABM(I,J),J=1,NREF)
      WRITE(6,560)
      DO 60 I=1,NG
   60 WRITE(6,570) I,(AS(I,J),J=1,NREF)
      IF(KZ.EQ.1) RETURN
```

```
      WRITE(6,580)
      DO 70 J=1,NREF
      WRITE(6,590)XREF(J),SHR2(J),AMM2(J)
  70  CONTINUE
 510  FORMAT(/,5X,'FINAL MOMENTS AND SHEARS')
 520  FORMAT(//5X,'MOMENTS IN GIRDERS:',/5X,20('*'),
     *          //5X,'REF. POINT NO.',12X,8(I6,4X))
 530  FORMAT(/5X,'DIST. FROM L.H.SUPPORT:',4X,8(F6.1,4X))
 540  FORMAT(/5X,'   GIRDER NO. ','                        MOMENTS ',
     *       /5X,80('-'))
 550  FORMAT(/5X,I4,21X,8(E10.4,3X))
 560  FORMAT(///5X,'CALCULATED SHEARS IN GIRDERS :',
     *          /5X,30('*'),/5X,'GIRDER NO.',27X,'SHEARS'//5X,83('-'))
 570  FORMAT(/5X,I5,20X,8(E10.4,3X))
 580  FORMAT(//5X,'          X              SHR              AMM',/)
 590  FORMAT(5X,3(F14.2,1X))
 600  FORMAT(1H1//5X,'OTPUT FOR HARMONICS NO. = ',I4)
      RETURN
      END
C
C

      SUBROUTINE DEFLEC(GMI,E,X1,I1,SPAN,W,DLS,WT,M)

      DIMENSION GMI(10),W(7),DLS(7)
      WT=0.0
      DO 30 I=1,M
      X=X1
      IF(X .G . DLS(I)) GOTO 10
      A=DLS(I)
      B=SPAN-A
      GOTO 20
  10  B=DLS(I)
      A=SPAN-B
      X=SPAN-X1
  20  WT=WT+W(I)*B*(A*SPAN*X+A*B*X-X*X*X)/(6*E*GMI(I1)*SPAN)
  30  CONTINUE
      RETURN
      END
C
C

      SUBROUTINE SOLVE(BI,NCOL,DI)
      DIMENSION BI(10,11),DI(10)

      N1=NCOL+1
      N2=NCOL-1
      DO 5 I=1,NCOL
   5  BI(I,N1)=DI(I)

      DO 40 K=1,N2
      K1=K+1
      DO 20 I=K1,NCOL
      P=BI(I,K)/BI(K,K)

      DO 10 J=K1,N1
      BI(I,J)=BI(I,J)-P*BI(K,J)

  10  CONTINUE
  20  CONTINUE
      DO 30 I=K1,NCOL
      BI(I,K)=0
```

```
   30 CONTINUE
   40 CONTINUE
      DI(NCOL)=BI(NCOL,N1)/BI(NCOL,NCOL)
      DO 60 NN=1,N2
      SUM=0
      I=NCOL-NN

      I1=I+1
      DO 50 J=I1,NCOL
      SUM=SUM+BI(I,J)*DI(J)

   50 CONTINUE
      DI(I)=(BI(I,N1)-SUM)/BI(I,I)

   60 CONTINUE
      RETURN
      END
C
C
      SUBROUTINE WDIST(N,NCOL,KGIR,XCOL,WB,M,W,DLS,CINF
     *,GMI,E,SPAN,HARC,BM,LCONT,WU,I3,KZ)
      DIMENSION KGIR(10),XCOL(10),WB(10),W(7),DLS(7),CINF(20,11),
     *GMI(10),HARC(5,20),WU(10,10),BM(5)
      PI=3.141592654
      IP=I3+1
      DO 10 I=1,NCOL
      I1=KGIR(I)
      IF (LCONT .EQ.1) CIN=CINF(I1,1)
      IF (LCONT .EQ.2) CIN=CINF(I1,IP)
      X=XCOL(I)
      CALL DEFLEC(GMI,E,X,I1,SPAN,W,DLS,WT,M)
      IF (KZ .EQ. 2) WRITE (6,510) WT,CIN
      YY=WT*CIN
      DO 10 IJ=1,N
      CONST1=IJ*PI*X/SPAN
      YY=YY-BM(IJ)*(CIN-HARC(IJ,I1))*SIN(CONST1)*SPAN*SPAN
     */(E*GMI(I1)*PI*PI*IJ*IJ)
      IF (LCONT .EQ. 1) WB(I)=YY
      IF (LCONT .EQ. 2) WU(I,I3)=YY
   10 CONTINUE
      IF (KZ .EQ.1) RETURN
      IF (LCONT .EQ. 1) WRITE (6,520)

      IF (LCONT .EQ. 2) WRITE(6,530)

      WRITE(6,540) (KGIR(I),I=1,NCOL)
      WRITE(6,550) (XCOL(I),I=1,NCOL)
      IF(LCONT .EQ. 2) GOTO 20
      WRITE (6,560) (WB(I),I=1,NCOL)

      RETURN
   20 CONTINUE
      DO 30 I=1,NCOL
      WRITE(6,570) (WU(I,J),J=1,NCOL)
   30 CONTINUE
  510 FORMAT (/,5X,'WDIST,WT,CIN',2E10.4)
  520 FORMAT (/5X,'DEFLECTIONS WHEN INTERMEDIATE SUPPORT
     *IS REMOVED')
  530 FORMAT(/5X,'MATRIX WU')
```

```
540 FORMAT(/5X,'GIRDER NO.',3X,10I6)
550 FORMAT(/5X,'DISTANCE FROM',/5X,'L.H. SUPPORT',10F6.1)
560 FORMAT(/5X,'DEFLECTION',/,10E10.4)
570 FORMAT(/5X,'DEFLECTION',3(3X,E10.4))
    RETURN
    END
C
C
    SUBROUTINE BEE(EF,WU,BI,NCOL)
    DIMENSION EF(10),WU(10,10),BI(10,11)
    DO 10 I=1,NCOL
    DO 10 J=1,NCOL
    BI(I,J)=WU(I,J)
    IF (I .EQ. J) BI(I,J)=BI(I,J)+EF(I)
 10 CONTINUE
    RETURN
    END
C
C
    SUBROUTINE DEE(WB,DELTA,DI,NCOL)
    DIMENSION WB(10),DELTA(10),DI(10)
    DO 10 I=1,NCOL
 10 DI(I)=WB(I)-DELTA(I)
    RETURN
    END
C
C
    SUBROUTINE EQN(I1,NG,AM,RM,XX,HARC,KZ,ICNK)
    DIMENSION S(20,21),B(20,21),F(20,21),T(20),XX(20),
   *  AM(20,20),RM(20),HARC(5,20)
    N2=0.0
    NG2=NG*2
    NG2U1=NG2+1
    DO 10 I=1,NG2
    XX(I)=0.
    T(I)=0.
    DO 10 J=1,NG2U1
    B(I,J)=0.
    F(I,J)=0.
 10     CONTINUE
    DO 20 I=1,NG2
    S(I,NG2U1)=RM(I)
    DO 20 J=1,NG2
    S(I,J)=AM(I,J)
 20     CONTINUE
    KI=NG2
    KJ=NG2U1
    DO 30  I=1,KI
    DO 30  J=1,KJ
 30     B(I,J)=S(I,J)
    N1=0
 40     CONTINUE
    N1=N1+1
    DO 50  I=1,KI
    DO 50  J=1,KJ
 50     F(I,J)=B(I,J)
    IF(N2.EQ.KI) GO TO 120
    DO 60 I=N1,KI
 60     T(I)=B(I,N1)
    N2=N1+1
```

```
          DO 80   I=N1,KI
          DO 80   J=N1,KJ
          IF(ABS(T(I)).EQ.0.0.OR.ABS(F(I,J)).EQ.0.) GO TO 70
          F(I,J)=F(I,J)/T(I)
          GO TO 80
70        F(I,J)=0.0
80        CONTINUE
          DO 90   I=N2,KI
          DO 90   J=N1,KJ
90        F(I,J)=F(N1,J)-F(I,J)
          DO 110 I=N2,KI
          DO 110 J=1,KJ
110       B(I,J)=F(I,J)
          GO TO 40
120       CONTINUE
          MB=KI
          NB=KJ
          DO 130 J=1,NB
130       B(1,J)=S(1,J)
          INB=NB-1
          IF(ABS(B(MB,NB)).EQ.0.0.OR.ABS(B(MB,MB)).EQ.0.0) GO TO 140
          XX(MB)=B(MB,NB)/B(MB,MB)
          GO TO 150
140       XX(MB)=0.0
150       CONTINUE
          N3=1
160       MB=KI
          MB=MB-N3
          KK=NB-N3
          TEXP=0.
          DO 170 J=KK,INB
170       TEXP=TEXP+XX(J)*B(MB,J)
          BWT=ABS(B(MB,NB)-TEXP)
          IF(BWT.EQ.0.0.OR.ABS(B(MB,MB)).EQ.0.0) GO TO 180
          XX(MB)=(B(MB,NB)-TEXP)/B(MB,MB)
          GO TO 190
 180      XX(MB)=0.0
 190      CONTINUE
          N3=N3+1
          IF(N3.EQ.KI) GO TO 200
          GO TO 160
200       CONTINUE
          IF(ICNK.EQ.1) GO TO 220
          DO 210 I=1,KI
210       HARC(I1,I)=XX(I)
220       CONTINUE
          IF(KZ.EQ.1) GO TO 250
          WRITE(6,510)
          DO 240 I=1,KI
          WRITE(6,520) I,XX(I)
240       CONTINUE
250       CONTINUE
          DO 260 I=1,20
          T(I)=0.
          DO 260 J=1,21
          B(I,J)=0.
          F(I,J)=0.
          S(I,J)=0.
260       CONTINUE
          N2=0
```

```
  510  FORMAT(//,15X,'CORRELATION COEFFICIENTS ',/15X,26('*'),//)
  520  FORMAT(15X,'B',I2,' = ',F14.5)
       RETURN
       END
C
       SUBROUTINE FINDEF(I3,NN,NCOL,NG,KGIR,XCOL,M,W,ARE,DLS,CINF,GMI,E,
      *SPAN,HARC,HARC1,BM,KZ,NREF,XREF)
       DIMENSION KGIR(10),XCOL(10),W(7),ARE(10),DLS(7),CINF(20,11),
      *          GMI(10),HARC(5,20),HARC1(5,20),BM(5),WF(10,10),W1(7),
      *          DLS1(7),NUM(10),XREF(10)
       IF(I3.GT.1)GOTO 35
       DO 10 I=1,10
       DO 10 J=1,10
   10  WF(I,J)=0.0
       PI=3.141592654
       DO 30 I=1,NREF
       DO 30 J=1,NG
       CIN=CINF(J,1)
       X=XREF(I)
       CALL DEFLEC(GMI,E,X,J,SPAN,W,DLS,WT,M)
       IF(KZ.EQ.2)WRITE(6,550)E,X,J,SPAN,WT,CIN
  550  FORMAT(/2X,' E ',F10.5,2X,' X ',F10.5,2X,' J',I3,2X,' SPAN',
      * F5.1,2X,' WT',F10.5,2X,'CIN ',F10.5)
       YY=WT*CIN
       DO 20 IJ=1,NN
       CONST1=IJ*PI*X/SPAN
   20  YY=YY-BM(IJ)*(CIN-HARC(IJ,J))*SIN(CONST1)*SPAN*SPAN/(E*GMI(J)*
      *PI*PI*IJ*IJ)
   30  WF(J,I)=YY
   35  CONTINUE
       W1(1)=-ARE(I3)
       DLS1(1)=XCOL(I3)
       M1=1
       CALL MOMENT(M1,NN,W1,DLS1,SPAN,BM,KZ)
       DO 60 J=1,NG
       DO 50 I=1,NREF
       IP=J+1
       I1=KGIR(I3)
       CIN=CINF(I1,IP)
       X=XREF(I)
       M1=1
       CALL DEFLEC(GMI,E,X,J,SPAN,W1,DLS1,WT,M1)
       YY=WT*CIN
       DO 40 IJ=1,NN
       CONST1=IJ*PI*X/SPAN
   40  YY=YY-BM(IJ)*(CIN-HARC1(IJ,J))*SIN(CONST1)*SPAN*SPAN/(E*GMI(J)*
      *PI*PI*IJ*IJ)
   50  WF(J,I)=WF(J,I)+YY
   60  CONTINUE
       IF(I3.NE.NCOL)RETURN
       DO 70 I=1,NREF
   70  NUM(I)=I
       WRITE(6,510) (NUM(I),I=1,NREF)
       WRITE(6,520) (XREF(I),I=1,NREF)
       WRITE(6,530)
       DO 80 I=1,NG
   80  WRITE(6,540) I,(WF(I,J),J=1,NREF)
```

```
510 FORMAT(/,5X,'FINAL DEFLECTION'//5X,'DEFLECTIONS IN GIRDERS:',
   */5X,20('*'),//5X,'REF. POINT NO.',12X,8(I6,4X))
520 FORMAT(/,5X,'DIST. FROM L.H. SUPPORT:',4X,8(F6.1,4X))
530 FORMAT(/,5X,'   GIRDER NO.',20X,'DEFLECTION',/,5X,80('-'))
540 FORMAT(/,5X,I4,21X,8(E10.4,1X))
    RETURN
    END
```

Index

ABOUT THE AUTHORS

The team of Jaeger and Bakht has been active in research in bridge engineering since 1976; its joint research has led to some 40 papers on various aspects of bridge engineering including analysis, testing, timber bridges, and probabilistic methods. One of their joint papers was recently awarded the Gzowski Medal by the Engineering Institute of Canada.

LESLIE G. JAEGER is a research professor of civil engineering and applied mathematics at the Technical University of Nova Scotia. A recipient of the Telford Premium from the Institution of Civil Engineers and of the A. B. Sanderson Award from the Canadian Society for Civil Engineering, and a fellow of the Royal Society of Edinburgh and the Canadian Society for Civil Engineering, he is the author of several books and numerous research publications. He is the technical editor of the Ontario Highway Bridge Design Code and also the chairman of its committee on methods of analysis. He obtained his Ph.D., and recently his D.Sc., from London University.

BAIDAR BAKHT is a principal research engineer and manager of the Structures Research Office at the Ministry of Transportation of Ontario. His research in the field of bridge engineering spans some 20 years, during which time he has contributed extensively to the technical literature. For his research publications he has been honored by awards from the American Society of Civil Engineers, the Canadian Society for Civil Engineering, the Engineering Institute of Canada, and the Road and Transportation Association of Canada. A fellow of the Canadian Society for Civil Engineering, he is the vice chairman of the Ontario Highway Bridge Design Code Committee and the chairman of its committee on wood structures. He obtained his M.Sc. from London University.